AF362038

Photonic Jet: Science and Application

Photonic Jet: Science and Application

Editors

Zengbo Wang
Boris Luk'yanchuk
Igor V. Minin

MDPI • Basel • Beijing • Wuhan • Barcelona • Belgrade • Manchester • Tokyo • Cluj • Tianjin

Editors
Zengbo Wang
School of Computer Science
and Electronic Engineering
Bangor University
Bangor
United Kingdom

Boris Luk'yanchuk
Faculty of Physics
Lomonosov Moscow State
University
Moscow
Russia

Igor V. Minin
Nondestructive Testing
School
Tomsk Polytechnic University
Tomsk
Russia

Editorial Office
MDPI
St. Alban-Anlage 66
4052 Basel, Switzerland

This is a reprint of articles from the Special Issue published online in the open access journal *Photonics* (ISSN 2304-6732) (available at: www.mdpi.com/journal/photonics/special_issues/pjsa).

For citation purposes, cite each article independently as indicated on the article page online and as indicated below:

LastName, A.A.; LastName, B.B.; LastName, C.C. Article Title. *Journal Name* **Year**, *Volume Number*, Page Range.

ISBN 978-3-0365-5272-9 (Hbk)
ISBN 978-3-0365-5271-2 (PDF)

Cover image courtesy of Zengbo Wang

Contents

About the Editors . . . . **vii**

Preface to "Photonic Jet: Science and Application" . . . . **ix**

Zengbo Wang, Boris Luk'yanchuk and Igor V. Minin
Special Issue on Photonic Jet: Science and Application
Reprinted from: *Photonics* **2022**, *9*, 540, doi:10.3390/photonics9080540 . . . . **1**

Heng Li, Wanying Song, Yanan Zhao, Qin Cao and Ahao Wen
Optical Trapping, Sensing, and Imaging by Photonic Nanojets
Reprinted from: *Photonics* **2021**, *8*, 434, doi:10.3390/photonics8100434 . . . . **7**

Fen Tang, Qingqing Shang, Songlin Yang, Ting Wang, Sorin Melinte and Chao Zuo et al.
Generation of Photonic Hooks from Patchy Microcylinders
Reprinted from: *Photonics* **2021**, *8*, 466, doi:10.3390/photonics8110466 . . . . **29**

Qingqing Shang, Fen Tang, Lingya Yu, Hamid Oubaha, Darwin Caina and Songlin Yang et al.
Super-Resolution Imaging with Patchy Microspheres
Reprinted from: *Photonics* **2021**, *8*, 513, doi:10.3390/photonics8110513 . . . . **39**

Rakesh Dhama, Bing Yan, Cristiano Palego and Zengbo Wang
Super-Resolution Imaging by Dielectric Superlenses: TiO$_2$ Metamaterial Superlens versus BaTiO$_3$ Superlens
Reprinted from: *Photonics* **2021**, *8*, 222, doi:10.3390/photonics8060222 . . . . **49**

Rayenne Boudoukha, Stephane Perrin, Assia Demagh, Paul Montgomery, Nacer-Eddine Demagh and Sylvain Lecler
Near- to Far-Field Coupling of Evanescent Waves by Glass Microspheres
Reprinted from: *Photonics* **2021**, *8*, 73, doi:10.3390/photonics8030073 . . . . **59**

Liyang Yue, Zengbo Wang, Bing Yan, Yao Xie, Yuri E. Geints and Oleg V. Minin et al.
Near-Field Light-Bending Photonic Switch: Physics of Switching Based on Three-Dimensional Poynting
Vector Analysis
Reprinted from: *Photonics* **2022**, *9*, 154, doi:10.3390/photonics9030154 . . . . **67**

Ksenia A. Sergeeva, Alexander A. Sergeev, Oleg V. Minin and Igor V. Minin
A Closer Look at Photonic Nanojets in Reflection Mode: Control of Standing Wave Modulation
Reprinted from: *Photonics* **2021**, *8*, 54, doi:10.3390/photonics8020054 . . . . **79**

Ching-Bin Lin, Yu-Hsiang Lin, Wei-Yu Chen and Cheng-Yang Liu
Photonic Nanojet Modulation Achieved by a Spider-Silk-Based Metal–Dielectric Dome Microlens
Reprinted from: *Photonics* **2021**, *8*, 334, doi:10.3390/photonics8080334 . . . . **89**

Djamila Bouaziz, Grégoire Chabrol, Assia Guessoum, Nacer-Eddine Demagh and Sylvain Lecler
Photonic Jet-Shaped Optical Fiber Tips versus Lensed Fibers
Reprinted from: *Photonics* **2021**, *8*, 373, doi:10.3390/photonics8090373 . . . . **99**

Hao Luo, Haibo Yu, Yangdong Wen, Jianchen Zheng, Xiaoduo Wang and Lianqing Liu
Direct Writing of Silicon Oxide Nanopatterns Using Photonic Nanojets
Reprinted from: *Photonics* **2021**, *8*, 152, doi:10.3390/photonics8050152 . . . . **107**

About the Editors

Zengbo Wang

Dr Zengbo Wang is the Head of the photonics and laser manufacturing group at Bangor University, United Kingdom. He received his Ph.D. from National University of Singapore and worked at University of Manchester and University of Southampton before he joined Bangor University in 2012. He is the recipient of the 2022 Leverhulme Trust Research Fellowship award. His research expertise lies in the fields of photonics, metamaterial and laser material processing, with special focuses on super-resolution science and technology, imaging, sensing, laser-based advanced manufacturing for industrial applications. He has published about 200 papers with pioneering contributions in microsphere super-resolution science and technology.

Boris Luk'yanchuk

Prof. Boris Luk'yanchuk graduated from Lomonosov Moscow State University, Russia, under the supervision of Prof. A. A. Abrikosov. Luk'yanchuk received his PhD from P.N. Lebedev Physical Institute, Academy of Sciences of USSR, 1979. Later, he worked as a Head of the Laboratory at General Physics Institute, Russian Academy of Sciences, Moscow. He is an Honorary Professor at Johannes Kepler University, Linz, Austria. During 1999-2019, Luk'yanchuk was working in the Data Storage Institute in Singapore, where he received the IES Prestigious Engineering Achievement Award 2004, President's Science Award 2013, and Institute of Physics Singapore (IPS) gold medal award 2016. The scientific interests of Boris Luk'yanchuk are related to laser–matter interactions. He has published 5 monographs and about 300 original papers.

Igor V. Minin

Prof. Igor V. Minin graduated from Novosibirsk State University, Russia. Igor Minin received his PhD from Leningrad University, 1986, and Dr. Sci (hub) from Novosibitsk State Technical University, 2004. He worked as a Head of the Laboratory at Institute of Applied Physics, Russian, Novosibirsk. He received the Rahmatulin Medal and award of the Russian Academy of Science. The scientific interests of Igor V. Minin are related to mesotronics (electromagnetic field–matter interactions at the mesoscale). He has published 11 monographs, more than 150 Russian patents and about 300 original papers, some of them (with Prof. O.V.Minin) are pioneering, including photonic hook.

Preface to "Photonic Jet: Science and Application"

Research on photonic jet science and technology started about two decades ago and has remained a very active field. However, there is no dedicated book or journal Special Issue published yet. We feel there is a need to edit a Special Issue on this topic to reflect the recent advances and developments as well as trends in the field. This Special Issue received many submissions and we selected ten papers of these to be included. Our editorial in this book provides an overview and detailed analysis of each work in the Special Issue. We hope you will find this book useful and valuable for photonic jet research.

Zengbo Wang, Boris Luk'yanchuk, and Igor V. Minin

Editors

Editorial

Special Issue on Photonic Jet: Science and Application

Zengbo Wang [1,*], Boris Luk'yanchuk [2] and Igor V. Minin [3]

[1] School of Computer Science and Electronic Engineering, Bangor University, Bangor LL57 1UT, UK
[2] Faculty of Physics, Lomonosov Moscow State University, 119991 Moscow, Russia; lukiyanchuk@nanolab.phys.msu.ru
[3] Nondestructive Testing School, Tomsk Polytechnic University, 36 Lenin Avenue, 634050 Tomsk, Russia; prof.minin@gmail.com
*   Correspondence: z.wang@bangor.ac.uk

Citation: Wang, Z.; Luk'yanchuk, B.; Minin, I.V. Special Issue on Photonic Jet: Science and Application. *Photonics* **2022**, *9*, 540. https://doi.org/10.3390/photonics9080540

Received: 28 July 2022
Accepted: 28 July 2022
Published: 1 August 2022

Publisher's Note: MDPI stays neutral with regard to jurisdictional claims in published maps and institutional affiliations.

Photonic jets (PJs) are important mesoscale optical phenomena arising from electromagnetic waves interacting with dielectric particles with sizes around several to several tens wavelengths (~2–40 λ). It generates a narrow, high-intensity beam at the shadow-side of the 'particle lenses' made from microspheres, cylinders (fibers), cubes, disks and others, even including biological cells and spider silks. A PJ has the capability to focus light beyond the classical diffraction limit, thus permitting many possibilities and applications: for example, super-resolution imaging, nanosensing, detection, patterning, trapping, manipulation, waveguiding, signal amplification (e.g., Raman, photoluminescence and second-harmonic generation) and high-efficiency signal collection, among others.

The earliest studies on PJ effects were reported in 2000 by Luk'yanchuk and co-workers [1,2]. They theoretically analyzed optical resonance and near-field effects [1], and verified experimentally [2]. By using 500 nm silica particles and a 248 nm-wavelength laser source, they obtained 100 nm hillocks ($\sim \lambda/2.5$) on a silicon surface [2]. Related works were carried by several groups between 2000 and 2004 [3,4]. Not knowing these works, Chen et al. reported in 2004 a theoretical subwavelength focusing effect by microcylinders and coined the terminology 'photonic nanojet' (PNJ) [5], which has since been widely used. A simplified term, 'photonic jet' (PJ), was also used by researchers in the field since 2005 [6,7]. In 2014, Minin et al. showed that a photonic jet can be formed from a three-dimensional particle of an arbitrary shape if the mesoscale condition is met [8,9]. In the past two decades, the field of PJ has undergone rapid growth and developments, driven by new innovations and discoveries. Among them, the most notable developments include the following: white-light microsphere nanoscope (2011) [10], spider silk superlens and metamaterial solid immersion lens (2016) [11,12], the discovery of photonic hooks (2016) [7,13], THz super-resolution imaging (2017) [14], single-cell biomagnifier (2019) [15], Plano-Convex-Microsphere (PCM) superlens (2020) [16,17], lipid droplets microlenses (2021) [18], PJ-mediated optogenetics (2022) [19] and others. More information on the past, present and future of PJ technology can be found in refs. [7,9,20–22].

This Special Issue focuses on the most recent advances and trends in PJ research. A total of 10 papers were selected and published, including a review of the field (one paper) [23], photonic hooks (three papers) [24–26], the modulation of PJ beam (three papers) [27–29], super-resolution imaging (three papers) [25,30,31], and scanning nanopatterning (one paper) [32]. Photonic hooks, field modulation and super-resolution imaging are the main topics in this Special Issue, which reflect the current research focus and trends. We highlight the key contribution and merit of the selected papers below according to the topics.

- Review on PJ-based trapping, sensing, and imaging

Li et al. reviewed the current types of microsphere lenses for PJ applications and their principles and applications in optical nano trapping, signal enhancement and super-resolution imaging, with particular emphasis on biological cells and tissues [23]. They envisage that in vivo nanomanipulation and biodetection will be the future trends. This review is an important addition to the existing literature on PJ technology, with refined focuses on trapping, sensing and imaging.

- Photonic hooks (PHs)

A PH is a new type of self-bending PJ beam, with a curvature that is less than the wavelength and differs from Airy-family beams [33]. Such structured beams were theoretically predicted in 2016 [7] and experimentally demonstrated in 2019 [34] based on dielectric Janus particles with broken symmetries in its geometry. Other methods for generating PH include asymmetric beam excitation and two material-composition particles, among others [22,33,35]. In the current issue, Tang et al. proposed a new method of using patchy microcylinders (i.e., partially metal-coated) for the generation of PHs [24] having in mind [35]. By rotating the patchy microcylinders, PHs with different curvatures were successfully demonstrated, and a bending angle of 28.4 degree and curved-line width of $0.36\ \lambda$ were reported. Based on this work, the authors extended their work to patchy microspheres and experimentally demonstrated, for the first time, the effect of PHs on super-resolution imaging [25]. They showed that PHs generated by patchy microspheres provide an effective oblique illumination of imaging objects, which leads to the considerable improvement of near-field imaging contrast, thus providing a new mean to optimize microsphere-based super-resolution techniques. On the other hand, Yue et al. demonstrated a new concept of using PHs for photonic switching [26,36]. When two-wavelengths beams, 1310 nm and 1550 nm, pass through a right-trapezoid dielectric Janus particle, they were separated and guided to different routes because of different bending strengths [26], thus permitting effective photonic switching on microscale through a simple dielectric particle. Potential applications in photonic integrated circuit are envisaged.

- Modulation of PJ beams

The ability to control and modulate PJ beams is essential for developing next-generation PJ devices and technologies. In principle, modulations can be achieved by various means, including tuning the particle's size, refractive index, surrounding medium, incident beam pupil and beam structure [37], the composition of particles (e.g., liquid crystals [38] and metamaterials [14,39]) and more. Here, Sergeeva et al. presented a new method for modulating PJs by using standing waves, which were achieved by positioning aluminum oxide hemispheres on top of a silicon substrate with a separation distance. The gap distance was chosen to match the phase conditions for constructive interference between incident and reflected beams [27] for effective modulation. The work offers a new pathway to design PJ-based integrated photonic devices. On the other hand, Lin et al. designed and manufactured a spider-silk-based metal-dielectric dome microlenses, which showed great performance in PJ modulations by using different metal casings [28]. When gold casing was used, the focusing intensity was maximized and increased by a factor of three due to the surface's plasmon resonance. This microlens could be used to scan a biological target for large-area imaging with a conventional microscope. In addition, Bouaziz et al. demonstrated another PJ modulation method based on fiber tip parameter tunning [29]. They showed that PJs were obtained when light is coupled in the guide's fundamental mode and when the base diameter of the microlens is close to the core's diameter and modulated by the sharpness of the tip. When the base diameter of the microlens is larger than the fiber core, the focus point tends to move away from the external surface of the fiber and has a larger width. The results of this study can be used as guidelines for the tailored fabrication of shaped optical-fiber tips according to the targeted application.

- Super-resolution imaging

Since the first demonstration of microsphere-assisted super-resolution imaging in 2011, the technology received wide attention and constant development by researchers across the world [21]. The field of development and roadmap has been systematically reviewed and summarized in refs. [20–22,40]. Note there are two types of particles superlenses used for super-resolution imaging in the literature: microsphere superlens [10] and metamaterial solid-immersion superlens [12], and both have different super-resolution imaging mechanisms. For microsphere superlens, the super-resolution mechanism is a combined contribution of several effects, including PJ focus, the excitation of super-resonance and whisper gallery modes, as well as substrate and partial and inclined illumination effects [21]. They lead to the conversion of high-frequency evanescent waves that contain near-field nanoscale information into propagating waves, which then reach the far-field and contribute to the formation of a magnified virtual image. Herein, by conducting rigorous electromagnetic simulations, Boudokha et al. provided new insight and evidence that microspheres are a natural collector and converter of evanescent waves to propagating waves using whisper gallery modes [30]. However, the evanescent-to-propagation-conversion (ETPC) efficiency can vary significantly (10^{-7} to 10^{-1}) depending on used microspheres. On the other hand, Dhama et al. designed and fabricated TiO_2 metamaterial superlens in full-sphere shape for the first time and compared its imaging performance with commonly used $BaTiO_3$ (BTG) microspheres under the same imaging settings [31]. Their results showed that the meta-superlens performs consistently better over the widely used BTG superlens in terms of imaging contrast, clarity, the field of view and resolution, which was further supported by theoretical simulation. In addition, as mentioned above, photonic hook (PH)-induced super contrast imaging was for the first time demonstrated by using a patchy microsphere [25] and asymmetric Janus particles [41]. These works will contribute to developments of more powerful, robust, and reliable super-resolution imaging systems, which have potential in revolutionizing the optical microscopy.

- Scanning nanopatterning

Alongside laser cleaning, surface nanopatterning are among the earliest applications of PJ effects. To fabricate arbitrary nanopatterns, various approaches such as angular beam scanning [42] and scanning Plano-convex-microsphere superlens [17] have been developed. Herein, Luo et. al. demonstrated a new laser-direct nanowriting system based on a combination of microsphere lens with an AFM cantilever and scanned over the sample's surface [31]. Using femtosecond laser sources, arbitrary silicon oxide nanopatterns with a feature size of 310 nm and height of 120 were directly fabricated in a single step. The proposed method shows the potential for the fabrication of multifunctional surfaces and silicon photonics and integrated chips.

We hope that this Special Issue will provide readers with a useful and timely update on the status and future trends of PJ research and mesotronics [7,20–22,33,43–47]. We thank all authors, reviewers and the *photonics* editorial team for their valuable contributions that brought this Special Issue to life.

Author Contributions: Paper writing: Z.W.; recommendations for invited papers and technical discussions about the content: Z.W., B.L. and I.V.M. All authors have read and agreed to the published version of the manuscript.

Funding: This research was funded by the Welsh European Funding Office (WEFO); European ERDF grants, grant number CPE 81400 and SPARCII c81133; UK Royal society (IEC\R2\202178 and IEC\NSFC\181378); Russian Science Foundation (no. 20-12-00389); Russian Foundation for Basic Research (no. 21-57-10001 and no. 20-02-00715); and the Tomsk Polytechnic University Development Program.

Institutional Review Board Statement: Not applicable.

Informed Consent Statement: Not applicable.

Data Availability Statement: Not applicable.

Conflicts of Interest: The authors declare no conflict of interest.

References

1. Luk'yanchuk, B.; Zheng, Y.W.; Lu, Y. Laser cleaning of solid surface: Optical resonance and near-field effects. *High-Power Laser Ablation III* **2000**, *4065*, 576–587.
2. Lu, Y.F.; Zhang, L.; Song, W.D.; Zheng, Y.W.; Luk'yanchuk, B.S. Laser writing of a subwavelength structure on silicon (100) surfaces with particle-enhanced optical irradiation. *JETP Lett.* **2000**, *72*, 457–459. [CrossRef]
3. Münzer, H.J.; Mosbacher, M.; Bertsch, M.; Zimmermann, J.; Leiderer, P.; Boneberg, J. Local field enhancement effects for nanostructuring of surfaces. *J. Microsc.* **2001**, *202*, 129–135. [CrossRef] [PubMed]
4. Chaoui, N.; Solis, J.; Afonso, C.N.; Fourrier, T.; Muehlberger, T.; Schrems, G.; Mosbacher, M.; Bäuerle, D.; Bertsch, M.; Leiderer, P. A high-sensitivity in situ optical diagnostic technique for laser cleaning of transparent substrates. *Appl. Phys. A* **2003**, *76*, 767–771. [CrossRef]
5. Chen, Z.; Taflove, A.; Backman, V. Photonic nanojet enhancement of backscattering of light by nanoparticles: A potential novel visible-light ultramicroscopy technique. *Opt. Express* **2004**, *12*, 1214. [CrossRef]
6. Lecler, S.; Takakura, Y.; Meyrueis, P. Properties of a three-dimensional photonic jet. *Opt. Lett.* **2005**, *30*, 2641–2643. [CrossRef]
7. Minin, I.V.; Minin, O.V. *Diffractive Optics and Nanophotonics: Resolution Below the Diffraction Limit*; Springer: Cham, Switzerland, 2016.
8. Minin, I.V.; Minin, O.V. Photonics of Isolated Dielectric Particles of Arbitrary Three-Dimensional Shape—A New Direction in Optical Information Technologies. *Vestnic NSU* **2014**, *12*, 4.
9. Minin, I.V.; Minin, O.V.; Geints, Y. Localized EM and photonic jets from non-spherical and non-symmetrical dielectric mesoscale objects: Brief review. *Ann. Phys.* **2015**, *527*, 491. [CrossRef]
10. Wang, Z.B.; Guo, W.; Li, L.; Luk'yanchuk, B.; Khan, A.; Liu, Z.; Chen, Z.C.; Hong, M.H. Optical virtual imaging at 50 nm lateral resolution with a white-light nanoscope. *Nat. Commun.* **2011**, *2*, 218. [CrossRef]
11. Monks, J.N.; Yan, B.; Hawkins, N.; Vollrath, F.; Wang, Z. Spider Silk: Mother Nature's Bio-Superlens. *Nano Lett.* **2016**, *16*, 5842–5845. [CrossRef]
12. Fan, W.; Yan, B.; Wang, Z.; Wu, L. Three-dimensional all-dielectric metamaterial solid immersion lens for subwavelength imaging at visible frequencies. *Sci. Adv.* **2016**, *2*, e1600901. [CrossRef]
13. Yue, L.Y.; Minin, O.V.; Wang, Z.B.; Monks, J.N.; Salin, A.S.; Minin, I.V. Photonic hook: A new curved light beam. *Opt. Lett.* **2018**, *43*, 771–774. [CrossRef]
14. Minin, I.V.; Minin, O.V. Terahertz artificial dielectric cuboid lens on substrate for super-resolution images. *Opt. Quantum Electron.* **2017**, *49*, 326. [CrossRef]
15. Li, Y.C.; Liu, X.S.; Li, B.J. Single-cell biomagnifier for optical nanoscopes and nanotweezers. *Light Sci. Appl.* **2019**, *8*, 61. [CrossRef]
16. Yan, B.; Song, Y.; Yang, X.B.; Xiong, D.X.; Wang, Z.B. Unibody microscope objective tipped with a microsphere: Design, fabrication, and application in subwavelength imaging. *Appl. Opt.* **2020**, *59*, 2641–2648. [CrossRef]
17. Yan, B.; Yue, L.Y.; Monks, J.N.; Yang, X.B.; Xiong, D.X.; Jiang, C.L.; Wang, Z.B. Superlensing plano-convex-microsphere (PCM) lens for direct laser nano-marking and beyond. *Opt. Lett.* **2020**, *45*, 1168–1171. [CrossRef]
18. Chen, X.X.; Wu, T.L.; Gong, Z.Y.; Guo, J.H.; Liu, X.S.; Zhang, Y.; Li, Y.C.; Ferraro, P.; Li, B.J. Lipid droplets as endogenous intracellular microlenses. *Light Sci. Appl.* **2021**, *10*, 242. [CrossRef]
19. Guo, J.H.; Wu, Y.; Gong, Z.Y.; Chen, X.X.; Cao, F.; Kala, S.; Qiu, Z.H.; Zhao, X.Y.; Chen, J.J.; He, D.M.; et al. Photonic Nanojet-Mediated Optogenetics. *Adv. Sci* **2022**, *9*, 2104140. [CrossRef]
20. Luk'yanchuk, B.S.; Paniagua-Domínguez, R.; Minin, I.V.; Minin, O.V.; Wang, Z. Refractive index less than two: Photonic nanojets yesterday, today and tomorrow. *Opt. Mater. Express* **2017**, *7*, 1820–1847. [CrossRef]
21. Wang, Z.; Luk'yanchuk, B. Super-resolution imaging and microscopy by dielectric particle-lenses. In *Label-Free Super-Resolution Microscopy*; Astratov, V., Ed.; Springer: Cham, Switzerland, 2019.
22. Minin, I.V.; Liu, C.Y.; Geints, Y.E.; Minin, O.V. Recent Advances in Integrated Photonic Jet-Based Photonics. *Photonics* **2020**, *7*, 41. [CrossRef]
23. Li, H.; Song, W.Y.; Zhao, Y.; Cao, Q.; Wen, A.H. Optical Trapping, Sensing, and Imaging by Photonic Nanojets. *Photonics* **2021**, *8*, 434. [CrossRef]
24. Tang, F.; Shang, Q.Q.; Yang, S.L.; Wang, T.; Melinte, S.; Zuo, C.; Ye, R. Generation of Photonic Hooks from Patchy Microcylinders. *Photonics* **2021**, *8*, 466. [CrossRef]
25. Qingqing, Q.Q.; Tang, F.; Yu, L.Y.; Oubaha, H.; Caina, D.; Yang, S.L.; Melinte, S.; Zuo, C.; Wang, Z.B.; Ye, R. Super-Resolution Imaging with Patchy Microspheres. *Photonics* **2021**, *8*, 513.
26. Yue, L.Y.; Wang, Z.B.; Yan, B.; Xie, Y.; Geints, Y.E.; Minin, O.V.; Minin, I.V. Near-Field Light-Bending Photonic Switch: Physics of Switching Based on Three-Dimensional Poynting Vector Analysis. *Photonics* **2022**, *9*, 154. [CrossRef]
27. Sergeeva, K.A.; Sergeev, A.A.; Minin, O.V.; Minin, I.V. A Closer Look at Photonic Nanojets in Reflection Mode: Control of Standing Wave Modulation. *Photonics* **2021**, *8*, 54. [CrossRef]
28. Lin, C.B.; Lin, Y.H.; Chen, W.Y.; Liu, C.Y. Photonic Nanojet Modulation Achieved by a Spider-Silk-Based Metal-Dielectric Dome Microlens. *Photonics* **2021**, *8*, 334. [CrossRef]
29. Bouaziz, D.; Chabrol, G.; Guessoum, A.; Demagh, N.E.; Lecler, S. Photonic Jet-Shaped Optical Fiber Tips versus Lensed Fibers. *Photonics* **2021**, *8*, 373. [CrossRef]

30. Boudoukha, R.; Perrin, S.; Demagh, A.; Montgomery, P.; Demagh, N.E.; Lecler, S. Near-to Far-Field Coupling of Evanescent Waves by Glass Microspheres. *Photonics* **2021**, *8*, 73. [CrossRef]
31. Dhama, R.; Yan, B.; Palego, C.; Wang, Z.B. Super-Resolution Imaging by Dielectric Superlenses: TiO_2 Metamaterial Superlens versus $BaTiO_3$ Superlens. *Photonics* **2021**, *8*, 222. [CrossRef]
32. Luo, H.; Yu, H.B.; Wen, Y.D.; Zheng, J.N.; Wang, X.D.; Liu, L.Q. Direct Writing of Silicon Oxide Nanopatterns Using Photonic Nanojets. *Photonics* **2021**, *8*, 152. [CrossRef]
33. Minin, O.V.; Minin, I.V. *The Photonic Hook: From Optics to Acoustics and Plasmonics*; Springer: Cham, Switzerland, 2021.
34. Minin, I.V.; Minin, O.V.; Katyba, G.M.; Chernomyrdin, N.V.; Kurlov, V.N.; Zaytsev, K.I.; Yue, L.; Wang, Z.; Christodoulides, D.N. Experimental observation of a photonic hook. *Appl. Phys. Lett.* **2019**, *114*, 031105. [CrossRef]
35. Minin, I.V.; Minin, O.V.; Liu, C.Y.; Wei, H.D.; Geints, Y.E.; Karabchevsky, A. Experiments demonstration of a tunable photonic hook by a partially illuminated dielectric microcylinder. *Opt. Lett.* **2020**, *45*, 4899–4902. [CrossRef]
36. Geints, Y.E.; Minin, O.V.; Yue, L.Y.; Minin, I.V. Wavelength-Scale Photonic Space Switch Proof-of-Concept Based on Photonic Hook Effect. *Ann. Phys.* **2021**, *533*, 2100192. [CrossRef]
37. Yue, L.Y.; Yan, B.; Monks, J.N.; Wang, Z.B.; Tung, N.T.; Lam, V.D.; Minin, O.V.; Minin, I.V. Production of photonic nanojets by using pupil-masked 3D dielectric cuboid. *J. Phys. D Appl. Phys.* **2017**, *50*, 175102. [CrossRef]
38. Eti, N.; Giden, I.H.; Hayran, Z.; Rezaei, B.; Kurt, H. Manipulation of photonic nanojet using liquid crystals for elliptical and circular core-shell variations. *J. Mod. Opt.* **2017**, *64*, 1566–1577. [CrossRef]
39. Yue, L.; Yan, B.; Wang, Z. Photonic nanojet of cylindrical metalens assembled by hexagonally arranged nanofibers for breaking the diffraction limit. *Opt. Lett.* **2016**, *41*, 1336–1339. [CrossRef]
40. Wang, Z.; Luk'yanchuk, B.; Wu, L. Roadmap for label-free imaging with microsphere superlens and metamaterial solid immersion lens. *arXiv* **2021**, arXiv:2108.09153.
41. Minin, O.V.; Minin, I.V. Terahertz microscopy with oblique subwavelength illumination in near field. *Quant. Electron.* **2022**, *52*, 13–16. [CrossRef]
42. Guo, W.; Wang, Z.B.; Li, L.; Whitehead, D.J.; Luk'yanchuk, B.S.; Liu, Z. Near-field laser parallel nanofabrication of arbitrary-shaped patterns. *Appl. Phys. Lett.* **2007**, *90*, 243101. [CrossRef]
43. Minin, I.V.; Minin, O.V.; Cao, Y.; Yan, B.; Wang, Z.; Luk'yanchuk, B. Photonic lenses with whispering gallery waves at Janus particles. *Opto-Electron. Sci.* **2022**, *1*, 21008. [CrossRef]
44. Zhang, P.C.; Yan, B.; Gu, G.Q.; Yu, Z.T.; Chen, X.; Wang, Z.B.; Yang, H. Localized photonic nanojet based sensing platform for highly efficient signal amplification and quantitative biosensing. *Sens. Actuators B-Chem.* **2022**, *357*, 131401. [CrossRef]
45. Luk'yanchuk, B.; Bekirov, A.R.; Wang, Z.B.; Minin, I.V.; Minin, O.V.; Fedyanin, A.A. Optical phenomena in dielectric spheres several light wavelengths in size: A review. *Phys. Wave Phenom.* **2022**, *30*, 217–241.
46. Minin, O.V.; Minin, I.V. Optical Phenomena in Mesoscale Dielectric Particles. *Photonics* **2021**, *8*, 591. [CrossRef]
47. Minin, I.V.; Minin, O.V.; Luk'yanchuk, B.S. Mesotronic era of dielectric photonics. *Mesophotonics Phys. Syst. Mesoscale* **2022**, *12152*, 89–93.

Review

Optical Trapping, Sensing, and Imaging by Photonic Nanojets

Heng Li, Wanying Song, Yanan Zhao, Qin Cao and Ahao Wen *

Institute of Nanophotonics, Jinan University, Guangzhou 511443, China; liheng@stu2019.jnu.edu.cn (H.L.);
songwanying2019@stu2019.jnu.edu.cn (W.S.); zhaoyanan@stu2020.jnu.edu.cn (Y.Z.); caoqin@jnu.edu.cn (Q.C.)
* Correspondence: wenahao3@jnu.edu.cn

Abstract: The optical trapping, sensing, and imaging of nanostructures and biological samples are research hotspots in the fields of biomedicine and nanophotonics. However, because of the diffraction limit of light, traditional optical tweezers and microscopy are difficult to use to trap and observe objects smaller than 200 nm. Near-field scanning probes, metamaterial superlenses, and photonic crystals have been designed to overcome the diffraction limit, and thus are used for nanoscale optical trapping, sensing, and imaging. Additionally, photonic nanojets that are simply generated by dielectric microspheres can break the diffraction limit and enhance optical forces, detection signals, and imaging resolution. In this review, we summarize the current types of microsphere lenses, as well as their principles and applications in nano-optical trapping, signal enhancement, and super-resolution imaging, with particular attention paid to research progress in photonic nanojets for the trapping, sensing, and imaging of biological cells and tissues.

Keywords: microspheres; photonic nanojets; optical trapping; signal enhancement; super-resolution imaging

Citation: Li, H.; Song, W.; Zhao, Y.; Cao, Q.; Wen, A. Optical Trapping, Sensing, and Imaging by Photonic Nanojets. *Photonics* **2021**, *8*, 434. https://doi.org/10.3390/ photonics8100434

Received: 30 August 2021
Accepted: 1 October 2021
Published: 11 October 2021

Publisher's Note: MDPI stays neutral with regard to jurisdictional claims in published maps and institutional affiliations.

1. Introduction

Optics are widely used in modern life and production, and are one of the frontier sciences in the field of modern science. Trapping [1–5], imaging [6–10], and sensing [11–15] nanoscale objects with light are becoming more important in nano-photonics and biomedicine. In 1590, Dutch eyeglass craftsman Hans Janssen and his son Zaccharias invented the first optical microscope, thereby breaking the limit of what the human eye could observe. With improvements by Leeuwenhoek and other scientists, optical microscopes have played a pivotal role in biological sciences, materials science, and other fields, leading to higher expectations for imaging resolution. In 1873, German physicist Abbe [16] discovered the formula for the resolution limit of a microscope: $d = \lambda/2n\sin\theta$, in which λ is the wavelength of illumination light, n is the refractive index of the imaged medium, and θ is the half angle at which the object receives the light from the object. When the illumination source is visible light, the resolution of the optical microscope is approximately 200 nm. In other words, if the distance between two points reaches 200 nm, it is not distinguishable by an optical microscope, which limits the development and progress of science. Similarly, the focused beam of optical tweezers also has a diffraction limit. Optical tweezers rely on a high numerical aperture. When particles with a diameter of less than 200 nm are in a focused beam, it is difficult to sense the gradient of light intensity. Therefore, it is difficult for optical tweezers to stably trap objects that are on the order of nanometers (<200 nm).

Researchers have recently made attempts to overcome the diffraction limit of traditional microscopes and optical tweezers [17–21]. Scanning electron microscopy (SEM) and transmission electron microscopy (TEM) have high resolution [22,23], and yet, the samples must be placed in a vacuum environment, which is not suitable for studying living biological samples. Fluorescence microscopy allows for the imaging of fluorescently labeled biological cell structures [24,25], enabling super-resolution imaging on the order of a few to several tens of nanometers. However, this technology requires the binding of fluorescent probes to the protein of the target sample, so only one type of protein can be imaged at a time. Additionally, the fluorescence intensity of the sample is easily extinguished over time. In 1972, Ash et al. obtained super-resolution microscopic images under an evanescent wave condition for the first time with a scanning near field microscope [26]. Soon after, researchers used a scanning near-field microscope with an optical probe tip close to the sample surface, scanning point-by-point in the near-field region, thus breaking the diffraction limit to obtain optical information on the sample surface [27,28]. At the same time, this approach takes a long time to obtain a complete image and it cannot be used to observe the sample in real time. To propagate the evanescent wave carrying sub-wavelength information to the far field, Pendry theoretically proposed a negative refractive index medium to enhance the evanescent wave to achieve sub-wavelength resolution, which provides the possibility to collect information in the far field [29]. Many researchers have used silver and other precious metal materials to prepare plasmon superlenses on this basis [30–33]. In recent years, plasmons have also been applied to the fields of optical trapping and optical manipulation [34–36]. When the wavelength of incident light irradiates the interface between the metal and medium, the free electrons on the metal surface oscillate. Resonance will occur when the wavelength of the incident light matches the resonant wavelength of the surface plasmon [37,38]. Under such resonance conditions, the energy of the electromagnetic field will be transformed into the collective vibrational energy of free electrons on the metal surface, and the light will be confined to the sub-wavelength range of the metal surface and be greatly enhanced. Similarly, photonic crystals were introduced to better break the diffraction limit and stably capture nanoparticles [39]. The coupling of the photonic crystal cavity and the laser makes the light intensity in the cavity increase, and the light force received by the nanoparticles becomes larger. However, for the optical trapping, imaging, and sensing of plasmon optical tweezers and photonic crystal optical tweezers, the absorption of light by the metal substrate can easily cause local thermal effects, thereby destroying the stability of the trap. More importantly, when the nanoparticles are biomaterials, the high temperature generated by the thermal effect will destroy the activity of biomolecules.

Compared with these complex technologies, microlens technology has been widely developed in the fields of super-resolution imaging [40], biosensing [41], and optical trapping [42] on the basis of its simplicity of preparation, ease of manipulation, and it being label-free. In 2004, the local photonic nanojet generated by the shadow surface of a micrometer-scale circular medium cylinder illuminated by a plane wave was first proved by Chen et al. By using high-resolution finite difference time domain (FDTD) numerical simulations, they found that the waist of the photonic nanojet is smaller than the diffraction limit, and it can propagate at multiple wavelengths without significant diffraction [43]. In 2011, Wang et al. [44] first reported microsphere lens nanoscopy that combines micron-scale transparent dielectric SiO_2 microspheres with conventional optical microscopy. The nanoscopy surpassed the diffraction limit under white light conditions to obtain optical imaging with 50 nm resolution. This simple and effective method can convert a near-field evanescent wave with high-frequency spatial information into propagation modes [45–47], offering the possibility to trap and detect nanoparticles [48–51], enhance the signal [52–55], mediate backaction force [56], and improve the performance of optical systems [57,58]. In this article, we will summarize the recent research progress of microsphere lenses, introduce three types of microsphere lenses, focus on the applications of microsphere lenses in optical trapping, sensing, and imaging, and discuss potential application scenarios.

2. Types and Principles of Microsphere Superlenses

2.1. Types of Microspheres

Microsphere superlenses can be classified by the medium in which the microspheres exist: microspheres in air medium, microspheres in liquid medium, and microspheres in solid medium.

In 2011, SiO_2 microspheres with a refractive index of 1.46 and diameter of 2–9 μm were directly placed on the surface of a sample by self-assembly technology to achieve super-resolution imaging of gold-plated oxide anodic alumina film with a spacing of 50 nm under a light source with a wavelength of 600 nm [44], as shown in Figure 1a. The microsphere lens allows for the collection of information about the object in the near field and the formation of a magnified virtual image in the far field. A resolution of $\lambda/8$–$\lambda/14$ and a magnification of $\times 8$ can be achieved in the air medium. In addition, according to theoretical calculations, the super-resolution intensity of microspheres with a refractive index of 1.8 was greatest in air. When the refractive index increases to 2.0, the super-resolution capability of microspheres becomes smaller. This demonstrates that not all microspheres have super-resolution capabilities; only microspheres that meet specific conditions can generate photonic nanojets to achieve super-resolution imaging. Microspheres in liquid media can be classified into two groups: semi-immersed in liquid and fully immersed in liquid. The experimental setup diagram is shown in Figure 1b [59]. Hao et al. showed that when SiO_2 microspheres with a refractive index of 1.47 and diameter of 3 μm were completely submerged in an ethanol solution, the microspheres did not have super-resolution capabilities [60]. When part of the ethanol solution was volatilized and the microspheres were semi-submerged in the solution, the contrast and resolution of the virtual image of the tested sample were enhanced, allowing imaging of commercial blue light discs with a width of 100 nm. When the ethanol solution was nearly evaporated and the microspheres were exposed to air, the resolution became weaker, further demonstrating that semi-immersion of the microspheres in liquid could improve the resolution of imaging. However, the volatility of the ethanol solution was not conducive to a prolonged observation of the experiment. Darafsheh et al. then demonstrated that super-resolution imaging can be achieved when high refractive index microspheres were completely submerged in a liquid solution [61,62]. Barium titanate ($BaTiO_3$) microspheres with a refractive index of 1.9 were completely submerged in an isopropyl alcohol solution with a refractive index of 1.37, and the super-resolution imaging of two-dimensional gold nanodimers comprising gold nanopillars with a diameter of 120 nm and height of 30 nm was achieved under an illumination light source with a wavelength of 550 nm. In 2014, the team demonstrated that the imaging effect $BaTiO_3$ microspheres with a refractive index of 2.1 submerged in an isopropane alcohol solution was better than that of soda lime glass with a refractive index of 1.51 in air [63]. Moreover, Darafsheh et al. proved that high refractive index microspheres embedded in a transparent film can achieve super-resolution imaging [64]. The experimental setup of microspheres in a solid medium is shown in Figure 1c. $BaTiO_3$ microspheres with a refractive index of 1.9 were self-assembled and distributed on liquid polydimethylsiloxane (PDMS). After drying and other steps to form PDMS films, the films embedded with microspheres were placed on the sample, and an illumination source of 550 nm was passed through the film to form a magnified virtual image below the specimen in the reflection illumination mode with a resolution of up to $\lambda/4$. Overall, microspheres with low refractive index can be directly imaged in air medium, and microspheres with high refractive index can be imaged at super-resolution in liquid and solid media. The microspheres in the liquid are flexible and can be trapped and manipulated with the help of auxiliary tools, such as optical fibers, to realize the imaging and sensing of specific particles and positions. The microspheres in the film can be prepared in advance and used as a cover glass for the sample, which can avoid the evaporation of liquid in the experiment and improve the stability of imaging.

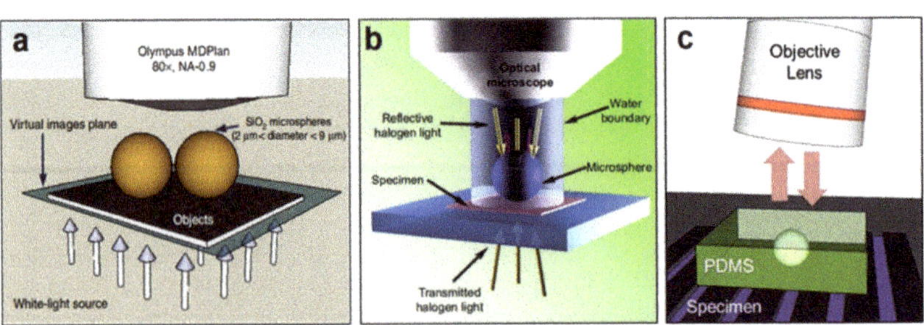

Figure 1. Schematic diagram of different microsphere superlens experimental setups. Schematic diagrams of the device with microspheres in an (**a**) air medium, (**b**) liquid medium, and (**c**) solid medium.

2.2. Principles of Photonic Nanojets for Optical Trapping, Sensing and Imaging

The optical properties of the photonic nanojets [65], whispering gallery mode [66], and directional antenna [67] of the microsphere lens enhance the interaction between photons and matter, thereby improving the ability of optical trapping, sensing, and imaging. The imaging of a conventional optical microscope comprises light spots formed by light and dark streaks. When two objects that are close pass through the lens, the diffracted light spots will overlap and be indistinguishable. Therefore, the resolution of conventional optical microscope depends on the size of the spot produced by focusing the incident light on the far field, which is limited by diffraction. For particle scales less than one-tenth of a wavelength, the scattered light intensity in each direction of the particle is not the same and is prone to Rayleigh scattering; the scattering cross section increases with the refractive index, and the electric field intensity increases with the refractive index in the near field range. When the particle scale is larger than the wavelength, Mie scattering will dominate and the microsphere will focus the incident beam into a photonic nanojet with a small width and high intensity [68,69]. The mechanism of the optical trapping, sensing, and imaging of microlens is usually explained by the photonic nanojet. The evanescent wave close to the surface of the microsphere plays an important role in the process of imaging the microsphere in conjunction with an optical microscope [70]. The sample is magnified by the microsphere lens, and then magnified twice by the microscope. The imaging mechanism of the microsphere lenses is shown in Figure 2a. The incident light can be considered a beam of parallel light when the incident waves irradiate the surface of the microsphere in the far-field range, and the distance between the sample and the wavelength of the light source are the same order of magnitude. When the parallel light passes through the surface of the microlens, it forms different angles of incidence, and then enters the interior through refraction. At this time, the evanescent wave carrying the high spatial frequency information of the object is converted into a propagating wave in the microsphere, which is received by a traditional optical microscope, and an enlarged virtual image can be produced in the far field [71]. In 2005, Li et al. demonstrated that the nano-photonic nanojet and backscattering capabilities of the microspheres were enhanced through FDTD simulation [72]. This shows that the microsphere lenses focus the light on the sub-diffraction limited size, realizes super-resolution imaging, and traps and senses particles to enhance the spectral signal [73,74]. Using an innovative 3D mapping technique, researchers have discovered significant field intensification around the poles of dielectric microspheres by tracking field-lines passing the critical points of the Poynting vector distribution [75].

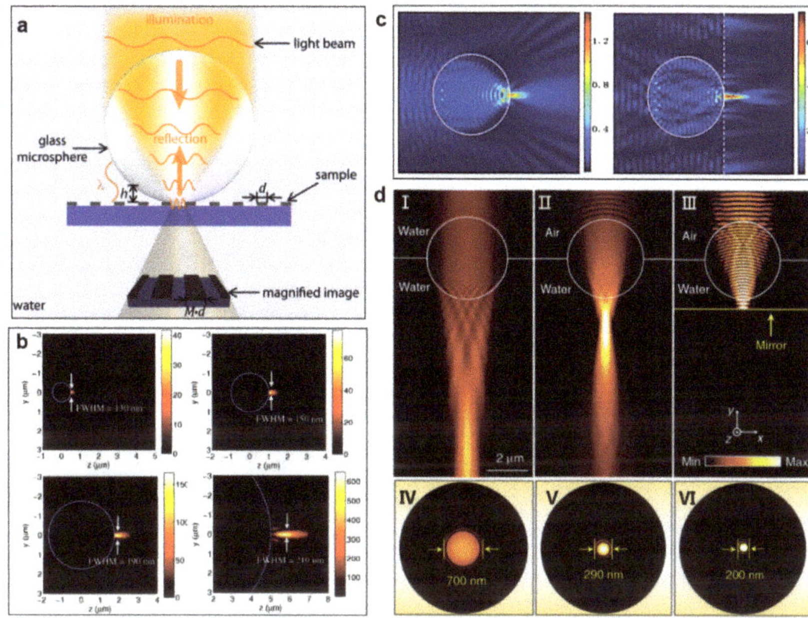

Figure 2. Principle of photonic nanojet generation. (**a**) Imaging schematic diagram of a grating sample using a microsphere superlens; (**b**) The near-field intensity and FWHM of the photonic nanojet generated by microspheres with diameters of 1.0 μm, 2.0 μm, 3.5 μm, and 8.0 μm; (**c**) Theoretical simulation of the light field distribution of microspheres on air and a Si substrate; (**d**) Optical intensity distributions of light focusing and light spots by a microlens fully immersed in water, semi-immersed in water, and suspended on the surface of a mirror.

The focusing ability of the photonic nanojet is related to the diameter and refractive index of the microspheres. Figure 2b shows the microsphere irradiated with a plane wave with a wavelength of 400 nm, in which the shadow side of the dielectric particles focuses a high-intensity photonic nanojet. Microspheres with different diameters were simulated by a finite element method. With the increase in diameter, the maximum light intensity and a full width at half maximum (FWHM) of the microspheres gradually increased; yet, the microspheres better convert evanescent waves into propagation waves the farther away the focus position is from the surface of the microspheres, and the high intensity energy carries more subwavelength information [72]. The refractive index of the microspheres also affects the imaging resolution. When the parallel light passes through the microspheres with high refractive index ($n > 2$), the focus is generated in the microspheres, so it is impossible to perform super-resolution imaging of the object. When the light passes through the microspheres with low refractive index ($1 < n < 2$), the focus is out of the microspheres, and the super-resolution imaging of the object can be performed [48]. Lee et al. also performed simulations using the Mie theory and found that the larger the refractive index of the microsphere [76], the stronger the intensity of the photonic nanojet. Additionally, the closer the strongest light intensity is to the surface of the microsphere, the higher the resolving power of the microsphere. The presence or absence of a substrate beneath the microsphere affects the formation of the shape and size of the photonic nanojets. As shown in Figure 2c, the incident light focuses near the shadow surface of the microsphere, forming a photonic nanojet with a focusing width of 150 nm. When the microsphere contacts the Si substrate, the electric field is redistributed around the sphere, and the formed photonic nanojet passes through the substrate with a focusing width of approximately 120 nm [77]. The properties of the photonic nanojet generated by the dielectric microspheres will change or even disappear for different substrate materials, leading to changes in the signal. In 2014, Sundaram's team further proved the importance of the substrate material on the imaging results by comparing the imaging resolution of microspheres selected without substrate, with aluminum oxide substrate materials and fused silica substrate materials [78]. As shown in Figure 2d, the intensity of the light field produced by the microlens is different in different environments [79]. When the 560 nm illuminating light is irradiated on the fully immersed microlens, the output light is focused on the far field with a focal length is 7.0 μm, and a relatively large output spot is produced. When the middle of the microlens is placed at the air–water interface, the output light is highly focused in the near field with a focal length of 0.7 μm, forming a tiny spot. When the distance between the microlens and the mirror substrate is 300 nm, the intensity of the light spot is generated by the light increases. This is due to the enhanced coherent interference between the photonic nanojet of the microlens and the reflected light from the specular surface.

In addition to spherical microlens, structures such as dielectric cubes, asymmetric cuboids, nanohole structured mesoscale dielectric spheres, and cylindrical objects can generate photonic nanojet, improve the spatial resolution of the imaging system, and even change the direction and focusing characteristics of the photonic nanojet to manipulate, sense, and image nanoscale objects [80–86]. The cuboid solid immersion lens can generate photonic nanojets though transmission and reflection modes to enhance the lateral resolution of the optical system [87]. When the dielectric cube is placed at the focus imaging point of the continuous wave terahertz imaging system or placed on the substrate, the spatial resolution of the imaging system can also be effectively improved [88,89]. Nguyen et al. placed a Teflon cube with a refractive index of 1.46 at the imaging point of the terahertz imaging system. After passing through the enhancer, the image contrast increased by a factor of 4.4. Besides, Ang et al. [90] attached a triangular prism to the irradiated surface of the cube. Due to the varying in the thickness of the prism, the phase of the transmitted waves in the upper and lower parts of the system changed, the electric field intensity became non-symmetric, showing concave deformation, which formed a curved photonic nanojet near the shadow surface. At the same time, the intensity of the photonic nanojet generated by the asymmetric cuboids was higher than that generated by cuboids and caused gold nanoparticles to move in a curved trajectory in the transmitted field (Figure 3a), to avoid obstacles. The shape and structure of the lens will also affect the length of the photonic nanojets. In recent years, researchers have changed the microstructure of the lens to obtain longer photonic nanojets. As shown in Figure 3b, Zhu et al. [91] obtained the ultra-long photonic nanojet by using the characteristics of the asymmetric two-microstructure formed by the support stage and the spherical cap. By appropriately adjusting the radius of curvature of the curved surface, an arbitrary elongated photonic nanojet can be obtained. Furthermore, the cascaded asymmetric silica microstructure will produce stable optical transmission and a FWHM waist close to $\lambda/4$. Gu et al. [92] used a plane wave to irradiate a liquid-filled hollow microcylinder to obtain the longest photonic nanojets. Immersion of the liquid-filled hollow cylinder into the solution environment can greatly spread the light beam. Because of the refractive index difference between the filling liquid and the immersion liquid, the focal length, attenuation length, and FWHM of the photonic nanojet can be flexibly adjusted by changing the inner filling liquid. Moreover, the permittivity contrast between the nanohole material and the dielectric particle results in the electric field enhancement of the nanohole-structured electric microspheres, which can produce high optical power and electric field intensity in the low refractive index hole material (air) [93–96]. In other words, a nanohole created on the back surface of a mesoscale particle in the medium can localize the field characteristics of the photonic nanojet to the size of the nanohole, thereby improving the resolution. As shown in Figure 3c, the focal spot size and focal volume of the nano-structured microspheres with a through hole of diameter $\lambda/5$ is larger than that without nanoholes. The larger nanoholes of $\lambda/5$ weaken the focusing ability of the dielectric microspheres. When the hole size is reduced by $\lambda/10$ or $\lambda/15$, the focal spot size and focal volume are significantly reduced. Similarly, dielectric microspheres with blind hole of diameter $\lambda/5$ on the surface, its focus ability are also weakened by the hole. When the aperture is reduced to below $\lambda/10$, the focused light spot is mainly limited to the blind hole.

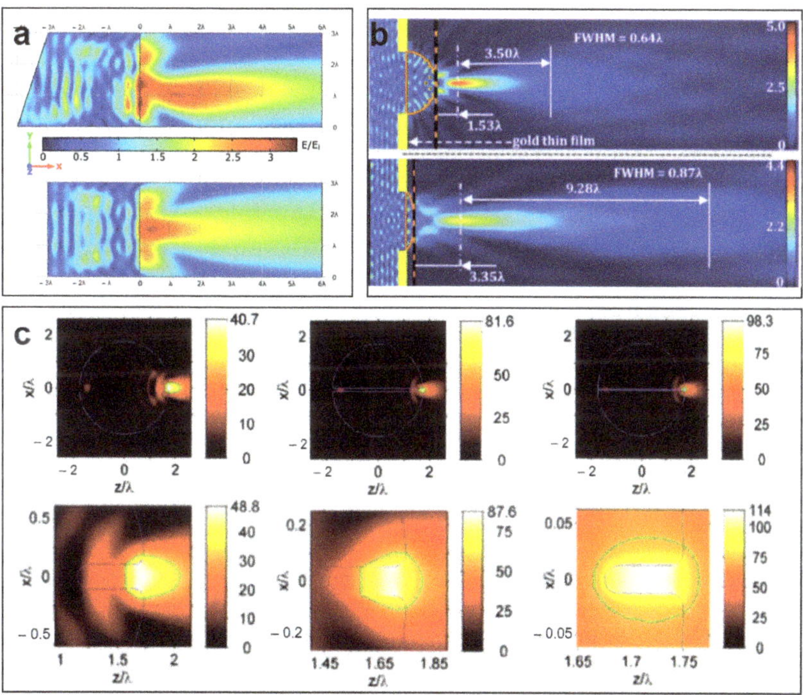

Figure 3. Generation of photonic nanojet of microlens with different structures. (**a**) Intensity of light field generated by asymmetric cuboids and cubes; (**b**) Spatial intensity distribution of structures with radius of curvature of 2λ and 4λ; (**c**) Light intensity of the simulated dielectric microspheres with a nanohole of size $\lambda/5$, $\lambda/10$, $\lambda/15$ and dielectric microspheres with a blind nanohole of the diameter $\lambda/5$, $\lambda/10$, $\lambda/40$.

Furthermore, the electromagnetic mode of objects also plays a crucial role in optical trapping, sensing, and imaging [97,98]. Using the whispering gallery mode can improve the imaging resolution and the possibility of particle detection. In the process of collecting point source information in whispering gallery mode, each external mode has an organized spatial spectrum, which can form an imaging mode with less than half of the illumination wavelength. When the object to be measured is close to the microsphere cavity, changes occur around the microsphere cavity, disturbing the whispering gallery mode. By receiving the resonant wavelength generated by the micro-cavity to obtain the change of the resonant frequency, the conversion of the evanescent wave to the propagating wave can be enhanced, and the corresponding external mode can be improved, so as to realize the detection of the refractive index of the surrounding environment or the concentration of molecules, as well as higher resolution imaging [99].

3. Optical Trapping and Sensing Using Photonic Nanojets

3.1. Fluorescence Signal Enhancement of Trapped Nano-Objects

Microsphere lenses can enhance the interaction of photons with matter under incident light irradiation, significantly enhancing the fluorescence signal [100,101] and sensing of the signal of manipulated objects in real time, providing a convenient method for nanomaterial characterization and biomolecular diagnosis. In 2015, Yang et al. probed the fluorescence signal of nanoparticles in microfluidic channels. When the nanoparticles pass through 3 μm melamine microspheres on a microcirculation channel, the photonic nanojets generated by the microsphere array are able to be transported in the flow medium and the fluorescence intensity of the nanoparticles is enhanced, such that the immune complexes formed on the Au nanoparticles can be detected [102].

However, most microsphere lenses cannot be adjusted and manipulated in the sample pool to detect objects. Lu et al. proved that the photonic nanojet generated by optical trapped microspheres can provide greater light power, making it easier to trap single 10 nm upconversion fluorescence nanoparticle (UCNP) [103]. The particles can be trapped and sensed by optical forces from fiber tweezers or by photophoresis [104–108]. As shown in Figure 4a, three-dimensional trapping and sensing the object can be implemented by combining optical fibers with microspheres [109]. Li et al. modified polystyrene (PS) microspheres or TiO_2 microspheres to adhere to the end face of negatively charged fiber tweezers. When trapping microlenses using fiber tweezers, the microlens generates a high-intensity photonic nanojet that manipulates the nanoparticles, which then acts as a high-value aperture objective for collecting the signal, and the fluorescent signal of the nanoparticles is enhanced when being sensed by the microlens adhered to the fiber tip. When sensing single nanoparticles in the presence of PS and TiO_2 microlenses, the fluorescence intensity of the trapped nanoparticles is 20 times and 30 times greater than the fluorescence intensity sensed by bare optical fibers, respectively. The excitation light passing through the microlens can produce a photonic nanojet phenomenon, in which the electric field intensity is enhanced in the local spot generated by the photonic nanojet, and this enhanced electric field contributes to the fluorescence excitation rate [110].

Dielectric microspheres act as microlenses to enhance fluorescence signals, and biological probes for the sensing and imaging of fluorescence signals from particles and biological tissues are also gradually being developed [111–113]. In 2017, Li et al. [114] used spherical yeast as a natural bio-microlens to enhance upconversion fluorescence, as shown in Figure 4b. The optical fiber is placed in the UCNPs. A laser with a wavelength of 980 nm and an optical power of 3 mW was emitted into the optical fiber. The fluorescence excited by the bare optical fiber was weak. The fluorescence intensity of the UCNPs was significantly enhanced when using fiber tweezers to trap the microlens. The use of a biological microlens can trap *Escherichia coli* (*E. coli*) and *Staphylococcus aureus* (*S. aureus*), which indicates that the presence of a biological microlens significantly enhances the upconversion fluorescence of *E. coli* and *S. aureus*. In addition, *S. aureus* and *E. coli* can be trapped and linked together, and their upconverted fluorescence signals can be simultaneously enhanced by approximately ~110. Moreover, Li et al. used living cells as biological lenses, demonstrating that cellular biological microlenses can also sense and enhance the fluorescence of particles with single-cell resolution [79]. The microlenses can also be manipulated in three dimensions by the light force generated by the optical tweezers. In 2020, using an optical tweezers system, Chen et al. moved $C_{10}H_7Br$ microlenses of different diameters above the CdSe@ZnS quantum dots with an emission wavelength of 550 nm [115]. The quantum dots were excited by the light of a mercury lamp filter. Under the microlens, the quantum dot fluorescence signal was sufficiently enhanced and detectable. By moving the microlens vertically along the Z axis, the brightest fluorescent spot in the field of view and the light intensity distribution corresponding to the dark field image were obtained, with a smaller diameter microlens boasting a strong signal enhancement (Figure 4c).

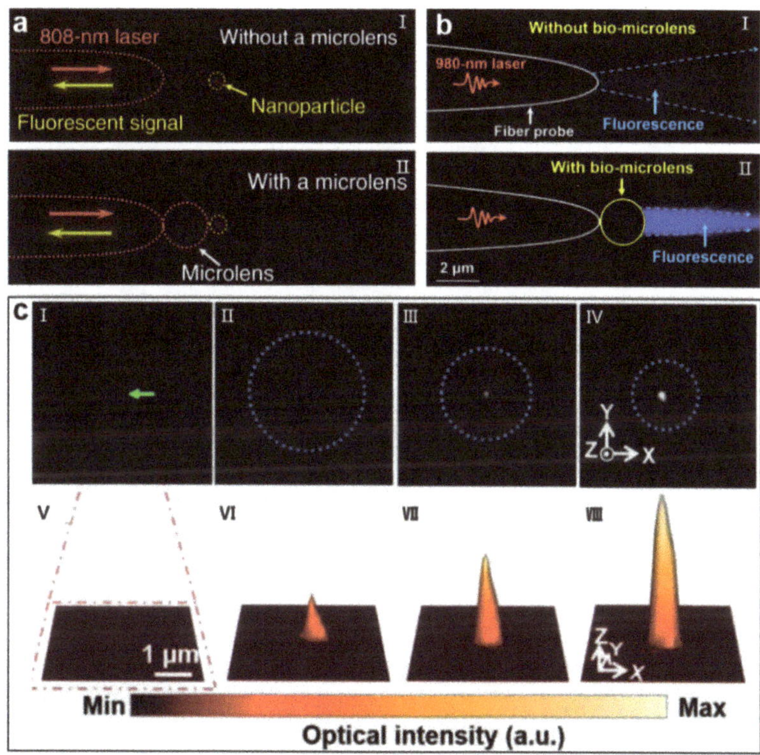

Figure 4. Fluorescence signal enhancement of microsphere superlenses. (**a**) Fluorescence signal images of the fiber without (I) and with (II) microlens for the sensing of individual nanoparticles; (**b**) Fluorescent image of the UCNP solution with fiber probe without (I) and with (II) biological microlens; (**c**) Fluorescence images of quantum dots with different diameters of $C_{10}H_7Br$ microlenses using optical tweezers.

3.2. Backscattering Signal Enhancement of Trapped Nano-Objects

When the highly focused beam generated by the microlens is irradiated on nanoparticles, the backscattering signal of the trapped nanoparticles can be significantly enhanced, thereby enhancing the sensing ability of the nanoparticles. It is beneficial to sense nanoparticles by analyzing the intensity and angular distribution of the enhanced backscattering from nanoparticles located in a photonic nanojet. In 2004, Chen et al. demonstrated through a two-dimensional numerical study that photonic nanojets can significantly enhance the backscattering of light by nanoparticles located within a nanojet [43]. Li et al. further proved enhanced backscattering of visible light by nanoparticles through a study of photonic nanojets [72]. Under light irradiation of different wavelengths, the backscattered signal of nanoparticles can be enhanced, whereas the enhanced backscattered power also varies, where dielectric microspheres act as microlenses to favor backscattered radiation [116]. Soon after, Yang et al. [117] experimentally verified for the first time that a photonic nanojet generated at visible wavelength can enhance the backscattering signal of nanoparticles. The photonic nanojet generated by $BaTiO_3$ microspheres with a diameter of 4.4 µm in the PDMS film can precisely locate and sense gold nanoparticle microspheres with diameters in the range of 50–100 nm.

As technology has evolved, there is considerable interest in high-resolution sensing systems that can trap and sense nano-objects and even single molecules in liquids. In 2015, researchers irradiated an array of melamine microspheres self-assembled in a microfluidic channel using an illuminated light source from an optical microscope, and the resulting photonic nanojets sensed 50–400 nm diameter Au nanoparticles flowing in the channel [102]. As shown in Figure 5a,b, the backscattered light intensity of Au nanoparticles at 200 nm on the photonic nanojet is approximately 40 times stronger than the backscattered light intensity outside the photonic nanojet. The backscattering signal of trapped nanoparticles can be sensed more flexibly by fiber tweezers with microlenses because of the small size of nanoparticles and their susceptibility to Brownian motion in solution. Li et al. used fiber tweezers to trap TiO_2 microlenses at the tip of a fiber probe [109], and a single 85 nm fluorescent nanoparticle was trapped and sensed by a highly focused photonic nanojet generated by a microlens. In the process of trapping fluorescent nanoparticles, the backscatter signal is divided into three processes: before trapping the nanoparticles, trapping the nanoparticles, and releasing the nanoparticles. When the nanoparticles are trapped, the intensity of the backscatter signal is significantly enhanced (Figure 5c). In addition, plasmid DNA biomolecules with low refractive index, small volume, and irregular shape can be sensed using the device because the photonic nanojet generated by the microlenses can enhance the backscattering signal.

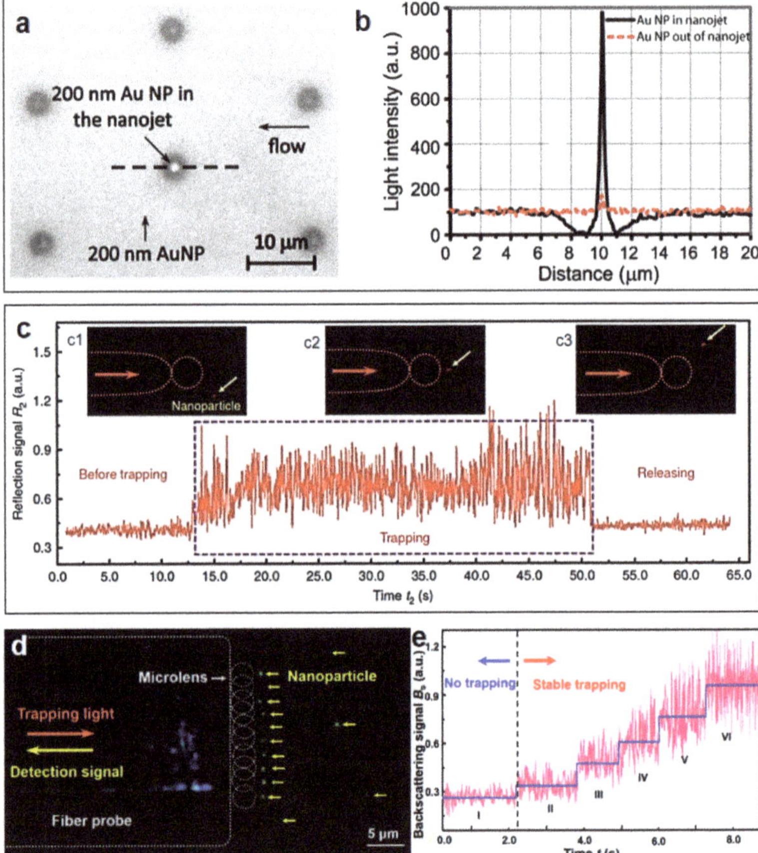

Figure 5. Backscattering signal enhancement of microlenses. (**a**) Two hundred nanometer diameter Au nanoparticles on a photonic nanojet; (**b**) Fluorescent image of UCNP solution with fiber probe without (I) and with (II) biological microlens; (**c**) Optical trapping of fluorescent nanoparticles by TiO_2 microlenses; (**d**) Optical images of fluorescent nanoparticles trapped by a microlens array; (**e**) Backscattering signals during trapping of multiple nanoparticles.

Next, Li's team assembled microsphere arrays on the end faces of fiber probes to trap and sense nanoparticles and subwavelength cells with high throughput, single nanoparticle resolution, and high selectivity [118]. As shown in Figure 5d,e, nanoparticles or cells were trapped using in-parallel photonic nanojet arrays, and their backscattered signals were sensing in real time with single-nanoparticle resolution, allowing for the detection of multiple nanoparticles and cells. To improve the sensitivity and biocompatibility of the detection, the team also used yeast as a biological microlens and trapped yeast using fiber tweezers to enhance the backscattering signal of *E. coli* chains [114], indicating prospects for single cell analysis and nanosensor applications.

3.3. Raman Signal Enhancement by Microsphere Superlens

Surface enhanced Raman scattering (SERS) is widely used in the analysis and sensing of materials. The Raman enhancement method of a photonic nanojet based on microspheres is a simple and reliable method. In 2007, Yi's team enhanced the Raman peak of Si by self-assembling SiO_2 microspheres on a silicon substrate because of the photonic nanojet effect produced by microspheres [119]. Transparent medium microspheres focus light to the finite size of sub-diffraction and focus visible light strongly in the photonic nanojet. As a result, the Raman signal of the measured object can be enhanced using microspheres [120]. In 2010, Du et al. demonstrated that a single dielectric microsphere can also enhance the Raman signal and that the enhancement is related to the size of the microsphere [77]. As shown in Figure 6a, a Raman peak was detected at 520 cm^{-1} when a PS microsphere with a refractive index of 1.59 was placed on the surface of a single crystal Si, while the Raman spectrum of only the PS microsphere had no peak at the same wavelength. This indicates that the characteristic peak of Si is significantly enhanced in the presence of a microlens.

In addition, a self-assembled high refractive index droplet microlens can enhance the Raman signal of Si wafers [115]. For bare silicon wafers or wafer regions without droplet microlenses, the detected Raman signal was very weak. When a suspension of the droplet microlens is placed on the silicon wafer, the microlens adheres to the silicon wafer surface by gravity, and the Raman signal of the silicon wafer is fully enhanced. The enhancement of the Raman signal is also different for droplet microlenses with different diameters (Figure 6b). The combination of a microsphere superlens and a solid film can also enhance the detection of Raman signals. Xing et al. immersed a monolayer of highly refractive $BaTiO_3$ microspheres into PDMS membranes and then transferred them to the sample surface for Raman detection [121]. As shown in Figure 6c,d, flexible microspheres embedded in thin films can enhance the Raman signal of one-dimensional carbon nanotubes and two-dimensional graphene. Furthermore, crystal violet molecules and Sudan I molecules can be tracked and sensed in aqueous solutions at a concentration of 10^{-7} M by coupling the flexible microsphere embedded film with silver nanoparticles or silver films. The flexible microsphere embedded film increases the SERS of the sample by 10 times and increases the sensing limit by at least an order of magnitude. To sense Raman signals more flexibly, microlenses can be combined with fiber probes [122]. Laser enhancement was achieved by focusing the incident laser on the silicon wafer surface through the microsphere on the probe, and it was observed experimentally that the tapered fiber could not effectively enhance the Raman scattering signal, and the Raman signal increased with the distance from the Raman microscope focal length.

All of the above Raman enhancement methods use fixed microspheres to enhance localized areas underneath them for single point acquisition of the sample. Recently, some researchers achieved Raman mapping enhancement of samples using microspheres [123]. As shown in Figure 6e,f, 5 μm SiO_2 microspheres attached to two vertical optical fibers were placed on a polysilicon substrate. As the sample is mapped, the microsphere stays under the laser beam of the objective lens and at the position of the microscope objective while the substrate moves below it. Therefore, all points of the image can be enhanced, and the signal enhancement at each point is $\times 4$.

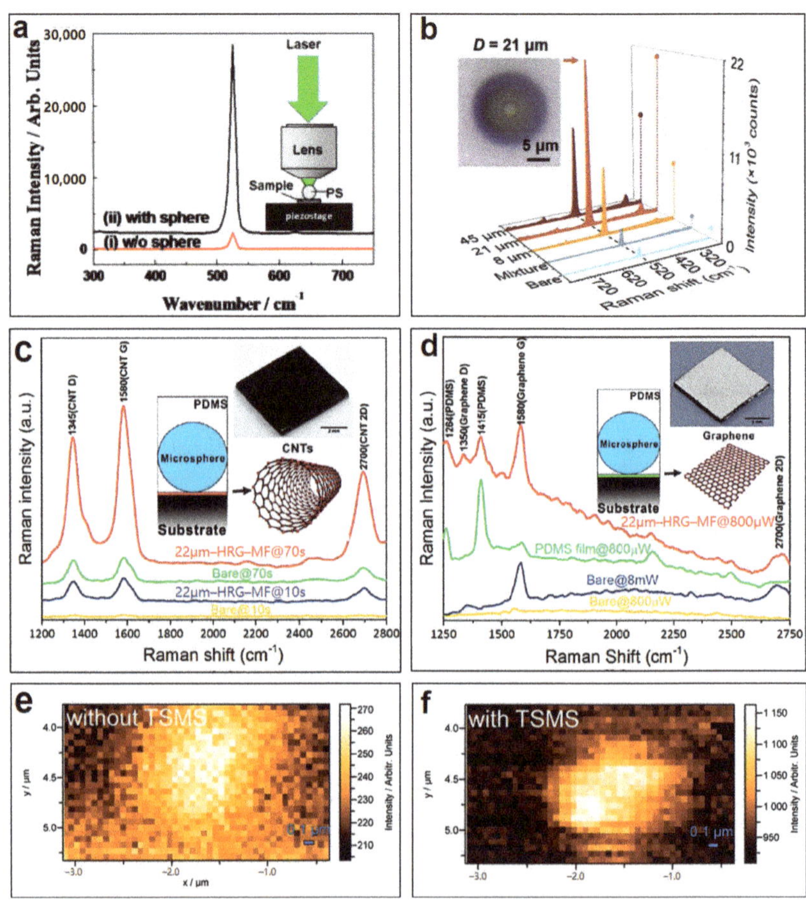

Figure 6. Raman enhancement of the microsphere lens. (**a**) Raman spectra of a Si wafer without PS microsphere (i) and with PS microsphere (ii); (**b**) Raman scattering intensity of different diameters of $C_{10}H_7Br$ microlenses on silicon wafers; Raman spectra of microspheres with and without high refractive index on (**c**) 1D carbon nanotubes and (**d**) 2D graphene; Raman mapping (**e**) with and (**f**) without microlens.

4. Super-Resolution Imaging by Photonic Nanojets

4.1. Optical Imaging of Nanostructures with Movable Microspheres

Microsphere lenses can be prepared by a variety of materials and methods [124–127], allowing high-resolution optical imaging of solid nanostructures at very low light intensities. As shown in Figure 7a, Wang et al. used SiO_2 microspheres with a diameter of 4.74 μm to image a gold-plated porous anodic aluminum oxide film with a diameter of 50 nm under white light conditions [44]. This method achieves real-time, label-free super-resolution imaging under white light conditions. Darafsheh et al. [63] immersed 5 μm $BaTiO_3$ microspheres in a liquid and achieved imaging of nanoplasmonic samples with a gap of 50–60 nm under irradiation at a wavelength of 405 nm (Figure 7b). In addition, super-resolution imaging of 250 nm metal fringes was achieved by the nano-scale flat spherical microlens prepared by chemically assembling the organic molecule hydroquinone from bottom to top [118]. Lee et al. used TiO_2 with a diameter of 60 μm and a refractive index of 2.2 to wrap ZnO, and the structure of 100–200 nm on Blu-ray discs was observed using a standard optical microscope [128]. Furthermore, Fan et al. [129] compactly stacked 45 nm anatase TiO_2 nanoparticles with a transparent refractive index of 2.55 using a solid-phase fluidic method. When a superlens comprising TiO_2 was located on a semiconductor wafer containing a parallel line pattern or a dotted line pattern, an image with a pitch of 60 nm and a complex structure of 50 nm was observed (Figure 7c). Dhama et al. [130] theoretically and experimentally demonstrated that a superlens comprising TiO_2 nanoparticles consistently outperformed $BaTiO_3$ microspheres in terms of imaging contrast, sharpness, field of view, and resolution because the tightly stacked 15 nm anatase TiO_2 nanoparticle composites have tiny air gaps between the particles, causing a dense scattering medium. Moreover, TiO_2 has almost no visible wavelength of energy dissipation. As a result, this near-field coupling effect between adjacent nanoparticles can be effectively propagated through the medium over long distances. The nanoparticle-synthesized medium will have the unusual ability to transform far-field illumination into large-area, nanoscale fading-wave illumination focused on the surface of an object in the near-field region. In addition, Wang et al. [131] used cylindrical spider silk under a traditional white light microscope with a wavelength of 600 nm to clearly distinguish 100 nm objects. This is due to the near-field interaction between the spider silk and the underlying nano-object, which causes the high spatial frequency evanescent wave at the surface boundary to be converted into a propagating wave. However, under dry conditions, super-resolution imaging cannot be achieved with spider silk. When isopropanol is used to fill regional gaps, the object can be super-resolution imaged due to the capillary binding force that occurs in the interface area. When the incident angle changes, the distance between the object and the lens also changes, so that the magnification factor can be adjusted.

To further increase the field of view of the microspheres in super-resolution imaging, large-area imaging can be achieved at a controllable position. Li et al. achieved stable and controllable image scanning of samples using chemical dynamics to drive the microsphere lens [132]. In addition, various attempts have been made to improve the field of view of microspheres in super-resolution imaging and achieve large-area imaging in a controllable position [133,134]. Krivitsky et al. achieved sample imaging of gold split squares deposited on silicon substrates with 73 nm gaps using a micropipette for accurate positioning between the squares [135], as shown in Figure 7d. The microsphere can also be combined with the cantilever of an atomic force microscope. The position of the microsphere can be changed by moving the cantilever, so that near-field information of the target position can be collected, and super-resolution images of any sample area can be obtained [136]. As shown in Figure 7e, the microspheres on the cantilever are used to approach the sample to realize imaging of the disc with a spacing of 80 nm. Moreover, the fiber probe can also act as a cantilever to improve the flexibility of imaging, using fiber tweezers to trap cells and scan the characters etched on the silicon substrate at a rate of ~20 μm/s [79], as shown in Figure 7f. Additionally, a 2×2 $C_{10}H_7Br$ droplet microlens array was assembled using optical tweezers [115] and the assembled droplet microlens was transferred to the polystyrene nanoparticle surface of the stack, where the contour of the nanoparticle became apparent in the field of view of the microscope (Figure 7g). Allen et al. [137] used high refractive index (n = 2) $BaTiO_3$ microspheres embedded in PDMS films to achieve large area imaging of 60 nm Au dimer spacing and 15 nm butterfly junction arrays. Zhang et al. [138] used $BaTiO_3$ microspheres embedded in PDMS films to image the streak structure on the surface of a Blu-ray disc (Figure 7h). Furthermore, through the dynamic scanning imaging mode of the microlens array and the superimposed reconstruction mode of the random microlens array area imaging, a 900 $μm^2$ surface image stitched by 210 images was realized (Figure 7i), which can reduce the number of images needed, improve imaging efficiency, and improve the observation range.

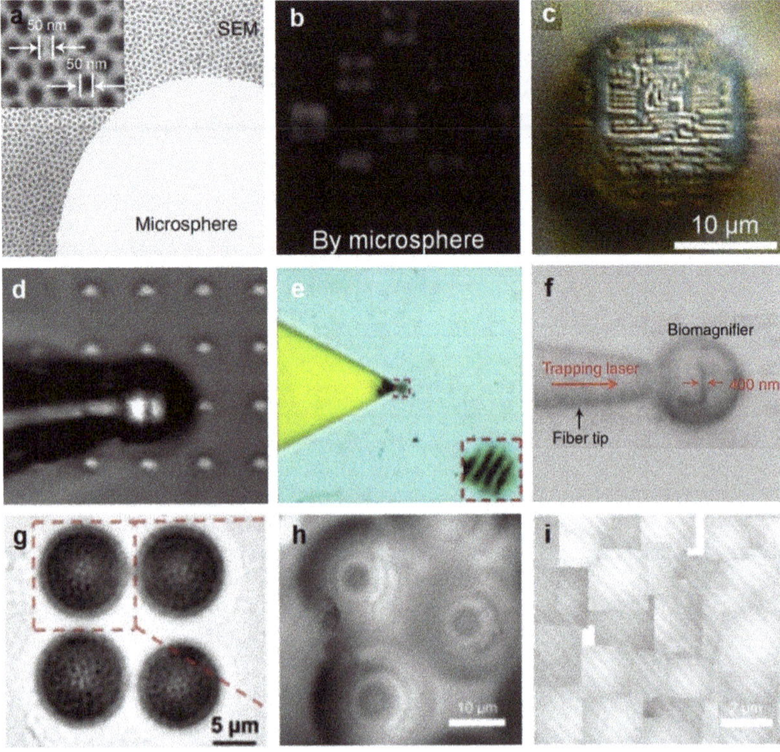

Figure 7. Optical imaging of nanostructures with microspheres. (**a**) SiO$_2$ microspheres on gold-plated porous anodic aluminum oxide film; (**b**) BaTiO$_3$ microspheres on nano-plasma samples with a gap of 50–60 nm; (**c**) TiO$_2$ microsphere superlenses on 60 nm wafers; (**d**) Magnified image of gold splitting square nanostructures imaged using microspheres combined with micropipettes; (**e**) Magnified image of a microsphere combined with an AFM cantilever against a DVD; (**f**) Optical images of nanopatterns trapped on the fiber of a biomagnifier; (**g**) Optical images of PS nanoparticles by a 2 × 2 microlens array; (**h**) Large-area imaging of Blu-ray discs by BaTiO$_3$ microlenses; (**i**) The Blu-ray disc surface recorded using the random microlens array area imaging superimposed reconstruction mode.

4.2. Super-Resolution Imaging of Living Cells by Photonic Nanojets

The combination of microsphere superlenses and an optical imaging device for biological imaging and analysis has been confirmed by many researchers. Fluorescence microscopy is generally used to observe cell structure and image bacteria, but the observation of specific biological organelles in vitro requires staining [139]. In 2014, Yang et al. [140] combined fluorescence microscopy with BaTiO$_3$ microspheres with a refractive index n = 1.92 and a diameter of 60 μm to image several different organelles in the alpha mouse liver 12 (AML12) cell line. As shown in Figure 8a, the traditional optical microscope can be used to identify the existence of centrioles, but it cannot observe their fine structure. The ring structure and γ-tubulin fluorescence labeling of the two centrioles was observed when the microsphere was placed above the centrosome, and even the junction of the two centrioles was identified. The mitochondria were then stained, and the fluorescence images of mitochondria were obtained using traditional light microscopy and microspheres. The complex shapes of the mitochondria were distinguished using BaTiO$_3$ microspheres. The influence of liquid evaporation on the imaging effect can be reduced by embedding the microspheres in transparent film and imaging with an inverted microscope. Darafsheh et al. obtained conventional fluorescence micrographs of cells under a fluorescence imaging mode by immersing 130 μm BaTiO$_3$ microspheres in PDMS films and placing them on the specimen under excitation at 365 nm [64]. At an excitation of 594 nm, the BaTiO$_3$ microspheres act as an auxiliary microlens to form a magnified virtual image below the sample surface. The red lesions of proton beam induced double-stranded DNA breaks were observed through the objective lens (Figure 8b).

In addition, the super-resolution imaging of certain organisms can be performed without fluorescent labeling. Li et al. used BaTiO$_3$ microspheres with a refractive index of 1.9 and diameter of 100 μm to image adenovirus at 75 nm under white light without fluorescent labeling or staining [59]. When BaTiO$_3$ microspheres were placed on the virus, the adenovirus could not be distinguished using a low magnification objective. When passed through a high-magnification objective, the microsphere converted the evanescent wave in the near field into an amplified extended wave in the far field through a frustrated total internal reflection mechanism, and individual adenoviruses were resolved and imaged below the sample surface (Figure 8c). The focusing of light by the microspheres forms a nanojet, which transmits the converted propagating waves to the space outside the sphere, which plays an important role in enhancing the image contrast. Furthermore, Chernomyrdin et al. used a method of the terahertz solid immersion microscope to focus electromagnetic waves to the evanescent field volume through a lens, thereby reducing the size of the terahertz beam caustics [141]. When articular hyaline cartilage of male sheep is placed on the scanning window of the system, the tissue ellipsoid of sub-wavelength diameter can be distinguished.

The optical imaging of subcellular structures is generally achieved by the combination of a microsphere lens and microscope. However, trapping the microlens with fiber tweezers can magnify and image any position of the biological sample in real time. As shown in Figure 8d, it is difficult to distinguish the bilayer structure of the fibrous cytoskeleton and cell membrane in the cell under an ordinary light microscope, and yet, after trapping the cell microlens with fiber tweezers and placing it above the epithelial cells, the bilayer structure of the fibrous cytoskeleton and cell membrane were clearly observed by enhancing the interaction between light and matter through the interference of incident and reflected light [79]. At the same time, Li et al. performed a numerical simulation analysis on the semi-immersed microlens above the mirror and obtained the focused photonic nanojet of the microlens. The theoretical imaging resolution is 85 nm, which is slightly smaller than the experimental resolution. This deviation is mainly due to the geometric configuration and refractive index of the cell microlens is defined as completely symmetrical and uniform in the simulation. In addition, Wang et al. proposed the use of atomic force microscopy in combination with microlenses to achieve large-area observations of living cell morphology or submembrane structures at sub-diffraction limited resolution [142]. As Figure 8e shows a fluorescent picture of C2C12 cells, the actin filaments in the cells can be clearly observed by moving the 56 μm diameter BaTiO$_3$ microspheres to the cell surface with the cantilever of an AFM, which is an important step for real-time monitoring of the status of a cell.

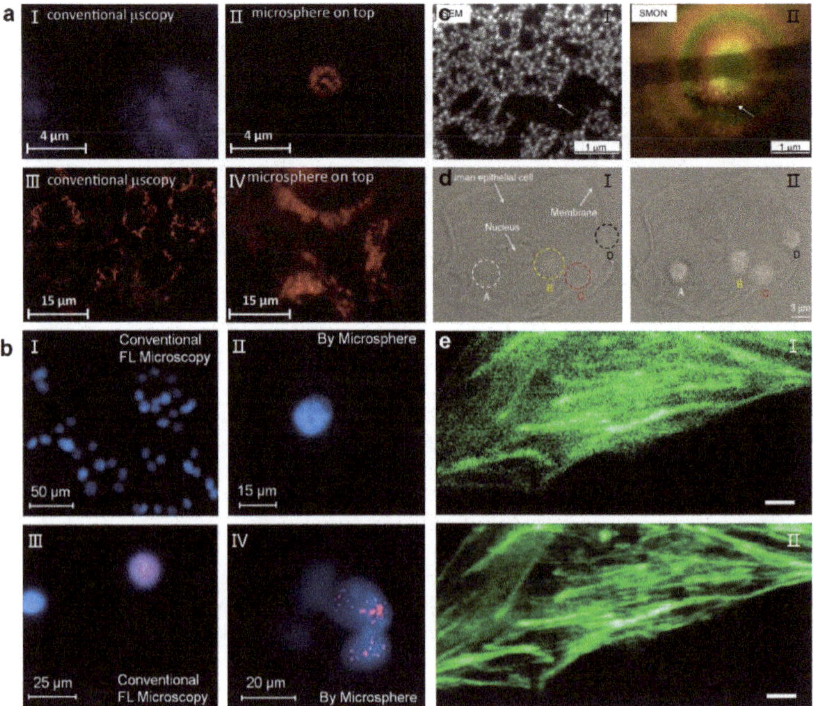

Figure 8. Super-resolution imaging of organisms. (**a**) Imaging of centrioles (I–II) and mitochondria (III–IV) of mouse hepatocyte lineage cells using conventional fluorescence microscopy and microsphere superlenses; (**b**) Optical images of cells without (I) and with (II) a microlens, and fluorescence images of cell nuclear double-stranded DNA imaging without a microlens (III) and with a microsphere superlens (IV); (**c**) SEM images of adenovirus by $BaTiO_3$ microspheres under white light (I) and imaging under $BaTiO_3$ (II).; (**d**) Imaging of the bilayer structure of the fibrous cytoskeleton and cell membrane without a microlens (I) and with a cellular lens (II); (**e**) Fluorescence imaging of C2C12 cells (I) and enhanced images with 56 μm diameter microsphere superlenses (II).

5. Conclusions and Outlook

This review systematically describes the application and progress of microsphere lenses in nano-optical trapping, sensing, and imaging from the types and principles of microsphere lenses. Thanks to the advantages of simple preparation, microsphere lenses provide a simple method for super-resolution imaging of biological samples and sensing of tiny particles, with potential value in biomedicine, microfluidics and nanophotonics. For another, microspheres can be combined with optical fibers, optical tweezers, and other tools to improve flexibility. Therefore, microspheres are expected to be constructed as photonic devices for biomedical imaging and real-time monitoring of samples, providing more promising technologies for biophotonics, nanophotonics, and biomedicine.

Most of the optical sensing and imaging based on microlenses are performed in vitro. However, these in vitro conditions cannot fully reflect the biological environment and conditions in vivo. Because the microlens is implantable, it has broad application prospects in in vivo nanomanipulation and biological detection. In addition, optical tweezers or optical traps provide a unique method of manipulating and controlling biological objects both in vivo and in vitro. The strong laser of light capture is prone to optical damage, which limits the exposure time of the captured sample. The photonic nanojet generated by the microlens may overcome optical opticution and allow the optical trapping of living cells to be widely implemented.

Author Contributions: Conceptualization, A.W.; writing—original draft preparation, H.L., W.S., and Y.Z.; writing—review and editing; A.W., Q.C. and H.L. All authors have read and agreed to the published version of the manuscript.

Funding: This research was funded by the Fundamental Research Funds for the Central Universities, grant number 21619321.

Institutional Review Board Statement: Not applicable.

Informed Consent Statement: Not applicable.

Data Availability Statement: The data that support the findings of this study and available from the corresponding author upon reasonable request.

Conflicts of Interest: The authors declare no conflict of interest.

References

1. Gu, M.; Bao, H.; Gan, X.; Stokes, N.; Wu, J. Tweezing and manipulating micro-and nanoparticles by optical nonlinear endoscopy. *Light Sci. Appl.* **2014**, *3*, e126. [CrossRef]
2. Gautam, R.; Xiang, Y.; Lamstein, J.; Liang, Y.; Bezryadina, A.; Liang, G.; Hansson, T.; Wetzel, B.; Preece, D.; White, A.; et al. Optical force-induced nonlinearity and self-guiding of light in human red blood cell suspensions. *Light Sci. Appl.* **2019**, *8*, 31. [CrossRef] [PubMed]
3. Bezryadina, A.S.; Preece, D.C.; Chen, J.C.; Chen, Z. Optical disassembly of cellular clusters by tunable 'tug-of-war' tweezers. *Light Sci. Appl.* **2016**, *5*, e16158. [CrossRef] [PubMed]
4. Singh, B.K.; Nagar, H.; Roichman, Y.; Arie, A. Particle manipulation beyond the diffraction limit using structured super-oscillating light beams. *Light Sci. Appl.* **2017**, *6*, e17050. [CrossRef] [PubMed]
5. Lepage, D.; Jiménez, A.; Beauvais, J.; Dubowski, J.J. Real-time detection of influenza a virus using semiconductor nanophotonics. *Light Sci. Appl.* **2013**, *2*, e62. [CrossRef]
6. Lee, S.; Kwon, O.; Jeon, M.; Song, J.; Shin, S.; Kim, H.; Jo, M.; Rim, T.; Dosh, J.; Kim, S.; et al. Super-resolution visible photoactivated atomic force microscopy. *Light Sci. Appl.* **2017**, *6*, e17080. [CrossRef]
7. Lombardini, A.; Mytskaniuk, V.; Sivankutty, S.; Andresen, E.R.; Chen, X.; Wenger, J.; Fabert, M.; Joly, N.; Louradour, F.; Kudlinski, A.; et al. High-resolution multimodal flexible coherent Raman endoscope. *Light Sci. Appl.* **2018**, *7*, 10. [CrossRef]
8. Petsch, S.; Schuhladen, S.; Dreesen, L.; Zappe, H. The engineered eyeball, a tunable imaging system using soft-matter micro-optics. *Light Sci. Appl.* **2016**, *5*, e16068. [CrossRef] [PubMed]
9. Zhang, W.; Zappe, H.; Seifert, A. Wafer-scale fabricated thermo-pneumatically tunable microlenses. *Light Sci. Appl.* **2014**, *3*, e145. [CrossRef]
10. Kim, K.; Jang, K.W.; Ryu, J.K.; Jeong, K.H. Biologically inspired ultrathin arrayed camera for high-contrast and high-resolution imaging. *Light Sci. Appl.* **2020**, *9*, 28. [CrossRef] [PubMed]
11. Li, Y.; Xin, H.; Zhang, Y.; Li, B. Optical fiber technologies for nanomanipulation and biodetection: A review. *J. Lightwave Technol.* **2021**, *39*, 251–262. [CrossRef]
12. Lei, H.; Zhang, Y.; Li, X.; Li, B. Photophoretic assembly and migration of dielectric particles and Escherichia coli in liquids using a subwavelength diameter optical fiber. *Lab Chip* **2011**, *11*, 2241–2246. [CrossRef]
13. Pan, T.; Lu, D.; Xin, H.; Li, B. Biophotonic probes for bio-detection and imaging. *Light Sci. Appl.* **2021**, *10*, 124. [CrossRef] [PubMed]
14. Su, J.; Goldberg, A.F.G.; Stoltz, B.M. Label-free detection of single nanoparticles and biological molecules using microtoroid optical resonators. *Light Sci. Appl.* **2016**, *5*, e16001. [CrossRef]
15. Yu, X.C.; Zhi, Y.; Tang, S.J.; Li, B.B.; Gong, Q.; Qiu, C.W.; Xiao, Y.F. Optically sizing single atmospheric particulates with a 10-nm resolution using a strong evanescent field. *Light Sci. Appl.* **2018**, *7*, 18003. [CrossRef] [PubMed]
16. Abbe, E. Beiträge zur Theorie des Mikroskops und der mikroskopischen Wahrnehmung. *Archiv für mikroskopische Anatomie* **1873**, *9*, 413–468. [CrossRef]
17. Hao, X.; Kuang, C.; Gu, Z.; Wang, Y.; Li, S.; Ku, Y.; Li, Y.; Ge, J.; Liu, X. From microscopy to nanoscopy via visible light. *Light Sci. Appl.* **2013**, *2*, e108. [CrossRef]
18. Yang, X.; Xie, H.; Alonas, E.; Liu, Y.; Chen, X.; Santangelo, P.J.; Ren, Q.; Xi, P.; Jin, D. Mirror-enhanced super-resolution microscopy. *Light Sci. Appl.* **2016**, *5*, e16134. [CrossRef] [PubMed]
19. Xiong, H.; Qian, N.; Miao, Y.; Zhao, Z.; Chen, C.; Min, W. Super-resolution vibrational microscopy by stimulated Raman excited fluorescence. *Light Sci. Appl.* **2021**, *10*, 87. [CrossRef]
20. Gao, D.; Ding, W.; Nieto-Vesperinas, M.; Ding, X.; Rahman, M.; Zhang, T.; Lim, C.; Qiu, C.W. Optical manipulation from the microscale to the nanoscale: Fundamentals, advances and prospects. *Light Sci. Appl.* **2017**, *6*, e17039. [CrossRef]

21. Yuan, G.H.; Rogers, E.T.F.; Zheludev, N.I. Achromatic super-oscillatory lenses with sub-wavelength focusing. *Light Sci. Appl.* **2017**, *6*, e17036. [CrossRef]
22. Eberle, A.L.; Mikula, S.; Schalek, R.; Lichtman, J.; Tate, M.K.; Zeidler, D. High-resolution, high-throughput imaging with a multibeam scanning electron microscope. *J. Microsc.* **2015**, *259*, 114–120. [CrossRef]
23. Browning, N.D.; Chisholm, M.; Pennycook, S.J. Atomic-resolution chemical analysis using a scanning transmission electron microscope. *Nature* **1993**, *366*, 143–146. [CrossRef]
24. Huang, B.; Bates, M.; Zhuang, X. Super-resolution fluorescence microscopy. *Annual Rev. Biochem.* **2009**, *78*, 993–1016. [CrossRef]
25. Dong, D.; Huang, X.; Li, L.; Mao, H.; Mo, Y.; Zhang, G.; Zhang, Z.; Shen, J.; Liu, W.; Wu, Z.; et al. Super-resolution fluorescence-assisted diffraction computational tomography reveals the three-dimensional landscape of the cellular organelle interactome. *Light Sci. Appl.* **2020**, *9*, 11. [CrossRef]
26. Ash, E.A.; Nicholls, G. Super-resolution aperture scanning microscope. *Nature* **1972**, *237*, 510–512. [CrossRef]
27. Pohl, D.W.; Denk, W.; Lanz, M. Optical stethoscopy: Image recording with resolution $\lambda/20$. *Appl. Phys. Lett.* **1984**, *44*, 651–653. [CrossRef]
28. Hosaka, N.; Saiki, T. Near-field fluorescence imaging of single molecules with a resolution in the range of 10 nm. *J. Microsc.* **2001**, *202*, 362–364. [CrossRef]
29. Pendry, J.B. Negative refraction makes a perfect lens. *Phys. Rev. Lett.* **2000**, *85*, 3966. [CrossRef]
30. Fang, N.; Lee, H.; Sun, C.; Zhang, X. Sub–diffraction-limited optical imaging with a silver superlens. *Science* **2005**, *308*, 534–537. [CrossRef]
31. Huang, S.; Wang, H.; Ding, K.H.; Tsang, L. Subwavelength imaging enhancement through a three-dimensional plasmon superlens with rough surface. *Opt. Lett.* **2012**, *37*, 1295–1297. [CrossRef]
32. Wang, C.; Zhang, W.; Zhao, Z.; Wang, Y.; Gao, P.; Luo, Y.; Luo, X. Plasmonic structures, materials and lenses for optical lithography beyond the diffraction limit: A review. *Micromachines* **2016**, *7*, 118. [CrossRef]
33. Barnes, W.L.; Dereux, A.; Ebbesen, T.W. Surface plasmon subwavelength optics. *Nature* **2003**, *424*, 824–830. [CrossRef]
34. Ivinskaya, A.; Petrov, M.I.; Bogdanov, A.A.; Shishkin, I.; Ginzburg, P.; Shalin, A.S. Plasmon-assisted optical trapping and anti-trapping. *Light Sci. Appl.* **2017**, *6*, e16258. [CrossRef]
35. Crozier, K.B. Quo vadis, plasmonic optical tweezers? *Light Sci. Appl.* **2019**, *8*, 35. [CrossRef]
36. Mestres, P.; Berthelot, J.; Aćimović, S.S.; Quidant, R. Unraveling the optomechanical nature of plasmonic trapping. *Light Sci. Appl.* **2016**, *5*, e16092. [CrossRef]
37. Li, Y.; Svitelskiy, O.V.; Maslov, A.V.; Carnegie, D.; Rafailov, E.; Astratov, V.N. Giant resonant light forces in microspherical photonics. *Light Sci. Appl.* **2013**, *2*, e64. [CrossRef]
38. Zhang, Y.; Min, C.; Dou, X.; Wang, X.; Urbach, H.P.; Somekh, M.G.; Yuan, X. Plasmonic tweezers: For nanoscale optical trapping and beyond. *Light Sci. Appl.* **2021**, *10*, 59. [CrossRef]
39. Bykov, D.S.; Xie, S.; Zeltner, R.; Machnev, A.; Wong, G.K.; Euser, T.G.; Russell, P.S.J. Long-range optical trapping and binding of microparticles in hollow-core photonic crystal fibre. *Light Sci. Appl.* **2018**, *7*, 22. [CrossRef]
40. Upputuri, P.K.; Pramanik, M. Microsphere-aided optical microscopy and its applications for super-resolution imaging. *Opt. Commun.* **2017**, *404*, 32–41. [CrossRef]
41. Toropov, N.; Cabello, G.; Serrano, M.P.; Gutha, R.R.; Rafti, M.; Vollmer, F. Review of biosensing with whispering-gallery mode lasers. *Light Sci. Appl.* **2021**, *10*, 42. [CrossRef]
42. Zhao, X.; Sun, Y.; Bu, J.; Zhu, S.; Yuan, X.C. Microlens-array-enabled on-chip optical trapping and sorting. *Appl. Opt.* **2011**, *50*, 318–322. [CrossRef]
43. Chen, Z.; Taflove, A.; Backman, V. Photonic nanojet enhancement of backscattering of light by nanoparticles: A potential novel visible-light ultramicroscopy technique. *Opt. Express* **2004**, *12*, 1214–1220. [CrossRef]
44. Wang, Z.; Guo, W.; Li, L.; Luk'Yanchuk, B.; Khan, A.; Liu, Z.; Chen, Z.; Hong, M. Optical virtual imaging at 50 nm lateral resolution with a white-light nanoscope. *Nat. Commun.* **2011**, *2*, 218. [CrossRef]
45. Ferrand, P.; Wenger, J.; Devilez, A.; Pianta, M.; Stout, B.; Bonod, N.; Popov, E.; Rigneault, H. Direct imaging of photonic nanojets. *Opt. Express* **2008**, *16*, 6930–6940. [CrossRef]
46. Yousefi, M.; Scharf, T.; Rossi, M. Photonic nanojet generation under converging and diverging beams. *J. Opt. Soc. Am. B* **2021**, *38*, 317–326. [CrossRef]
47. Han, L.; Han, Y.; Gouesbet, G.; Wang, J.; Gréhan, G. Photonic jet generated by spheroidal particle with Gaussian-beam illumination. *J. Opt. Soc. Am. B* **2014**, *31*, 1476–1483. [CrossRef]
48. Luk'yanchuk, B.S.; Paniagua-Domínguez, R.; Minin, I.; Minin, O.; Wang, Z. Refractive index less than two: Photonic nanojets yesterday, today and tomorrow. *Opt. Mater. Express* **2017**, *7*, 1820–1847. [CrossRef]
49. Darafsheh, A.; Bollinger, D. Systematic study of the characteristics of the photonic nanojets formed by dielectric microcylinders. *Opt. Commun.* **2017**, *402*, 270–275. [CrossRef]
50. Gu, G.; Song, J.; Chen, M.; Peng, X.; Liang, H.; Qu, J. Single nanoparticle detection using a photonic nanojet. *Nanoscale* **2018**, *10*, 14182–14189. [CrossRef]

51. Tsai, M.C.; Tsai, T.L.; Shieh, D.B.; Chiu, H.T.; Lee, C.Y. Detecting HER2 on cancer cells by TiO$_2$ spheres Mie scattering. *Anal. Chem.* **2009**, *81*, 7590–7596. [CrossRef]
52. Darafsheh, A. Photonic nanojets and their applications. *J. Phys. Photon.* **2021**, *3*, 022001. [CrossRef]
53. Lecler, S.; Haacke, S.; Lecong, N.; Crégut, O.; Rehspringer, J.L.; Hirlimann, C. Photonic jet driven non-linear optics: Example of two-photon fluorescence enhancement by dielectric microspheres. *Opt. Express* **2007**, *15*, 4935–4942. [CrossRef]
54. Biccari, F.; Hamilton, T.; Ristori, A.; Sanguinetti, S.; Bietti, S.; Gurioli, M.; Mohseni, H. Quantum dots luminescence collection enhancement and nanoscopy by dielectric microspheres. *Part. Part. Syst. Charact.* **2020**, *37*, 1900431. [CrossRef]
55. Das, G.M.; Ringne, A.B.; Dantham, V.R.; Easwaran, R.K.; Laha, R. Numerical investigations on photonic nanojet mediated surface enhanced Raman scattering and fluorescence techniques. *Opt. Express* **2017**, *25*, 19822–19831. [CrossRef]
56. Ren, Y.X.; Zeng, X.; Zhou, L.M.; Kong, C.; Mao, H.; Qiu, C.W.; Tsia, K.K.; Wong, K.K. Photonic nanojet mediated backaction of dielectric microparticles. *ACS Photonics* **2020**, *7*, 1483–1490. [CrossRef]
57. Sun, S.; Li, M.; Du, Q.; Png, C.E.; Bai, P. Metal–dielectric hybrid dimer nanoantenna: Coupling between surface plasmons and dielectric resonances for fluorescence enhancement. *J. Phys. Chem. C* **2017**, *121*, 12871–12884. [CrossRef]
58. Migliozzi, D.; Gijs, M.A.; Huszka, G. Microsphere-mediated optical contrast tuning for designing imaging systems with adjustable resolution gain. *Sci. Reports* **2018**, *8*, 15211. [CrossRef]
59. Li, L.; Guo, W.; Yan, Y.; Lee, S.; Wang, T. Label-free super-resolution imaging of adenoviruses by submerged microsphere optical nanoscopy. *Light Sci. Appl.* **2013**, *2*, e104. [CrossRef]
60. Hao, X.; Kuang, C.; Liu, X.; Zhang, H.; Li, Y. Microsphere based microscope with optical super-resolution capability. *Appl. Phys. Lett.* **2011**, *99*, 203102. [CrossRef]
61. Darafsheh, A.; Walsh, G.F.; Dal Negro, L.; Astratov, V.N. Optical super-resolution by high-index liquid-immersed microspheres. *Appl. Phys. Lett.* **2012**, *101*, 141128. [CrossRef]
62. Darafsheh, A.; Limberopoulos, N.I.; Derov, J.S.; Walker, D.E., Jr.; Durska, M.; Krizhanovskii, D.N.; Whittaker, D.M.; Astratov, V.N. Optical microscopy with super-resolution by liquid-immersed high-index microspheres. *Proc. SPIE* **2013**, *8594*, 85940C. [CrossRef]
63. Darafsheh, A.; Limberopoulos, N.I.; Derov, J.S.; Walker, D.E., Jr.; Astratov, V.N. Advantages of microsphere-assisted super-resolution imaging technique over solid immersion lens and confocal microscopies. *Appl. Phys. Lett.* **2014**, *104*, 061117. [CrossRef]
64. Darafsheh, A.; Guardiola, C.; Palovcak, A.; Finlay, J.C.; Cárabe, A. Optical super-resolution imaging by high-index microspheres embedded in elastomers. *Opt. Letters* **2015**, *40*, 5–8. [CrossRef] [PubMed]
65. Heifetz, A.; Kong, S.C.; Sahakian, A.V.; Taflove, A.; Backman, V. Photonic nanojets. *J. Comput. Theor. Nanosci.* **2009**, *6*, 1979–1992. [CrossRef]
66. Zhou, S.; Deng, Y.; Zhou, W.; Yu, M.; Urbach, H.P.; Wu, Y. Effects of whispering gallery mode in microsphere super-resolution imaging. *Appl. Phys. B* **2017**, *123*, 236. [CrossRef]
67. Ristori, A.; Hamilton, T.; Toliopoulos, D.; Felici, M.; Pettinari, G.; Sanguinetti, S.; Gurioli, M.; Mohseni, H.; Biccari, F. Photonic jet writing of quantum dots self-aligned to dielectric microspheres. *Adv. Quantum Technol.* **2021**, *4*, 2100045. [CrossRef]
68. Fan, X.; Zheng, W.; Singh, D.J. Light scattering and surface plasmons on small spherical particles. *Light Sci. Appl.* **2014**, *3*, e179. [CrossRef]
69. Kim, M.S.; Scharf, T.; Mühlig, S.; Rockstuhl, C.; Herzig, H.P. Engineering photonic nanojets. *Opt. Express* **2011**, *19*, 10206–10220. [CrossRef]
70. Wang, B.; Barbiero, M.; Zhang, Q.M.; Gu, M. Super-resolution optical microscope: Principle, instrumentation, and application. *Front. Inf. Technol. Electron. Eng.* **2019**, *20*, 608–630. [CrossRef]
71. Yang, H.; Trouillon, R.; Huszka, G.; Gijs, M.A. Super-resolution imaging of a dielectric microsphere is governed by the waist of its photonic nanojet. *Nano Lett.* **2016**, *16*, 4862–4870. [CrossRef]
72. Li, X.; Chen, Z.; Taflove, A.; Backman, V. Optical analysis of nanoparticles via enhanced backscattering facilitated by 3-D photonic nanojets. *Opt. Express* **2005**, *13*, 526–533. [CrossRef] [PubMed]
73. Kasim, J.; Ting, Y.; Meng, Y.Y.; Ping, L.J.; See, A.; Jong, L.L.; Xiang, S.Z. Near-field Raman imaging using optically trapped dielectric microsphere. *Opt. Express* **2008**, *16*, 7976–7984. [CrossRef] [PubMed]
74. Lecler, S.; Takakura, Y.; Meyrueis, P. Properties of a three-dimensional photonic jet. *Opt. Lett.* **2005**, *30*, 2641–2643. [CrossRef]
75. Yue, L.; Yan, B.; Monks, J.N.; Dhama, R.; Jiang, C.; Minin, O.V.; Minin, I.V.; Wang, Z. Full three-dimensional Poynting vector flow analysis of great field-intensity enhancement in specifically sized spherical-particles. *Sci. Rep.* **2019**, *9*, 20224. [CrossRef] [PubMed]
76. Lee, S.; Li, L.; Wang, Z.; Guo, W.; Yan, Y.; Wang, T. Immersed transparent microsphere magnifying sub-diffraction-limited objects. *Appl. Opt.* **2013**, *52*, 7265–7270. [CrossRef] [PubMed]
77. Du, C.; Kasim, J.; You, Y.M.; Shi, D.N.; Shen, Z.X. Enhancement of Raman scattering by individual dielectric microspheres. *J. Raman Spectrosc.* **2011**, *42*, 145–148. [CrossRef]
78. Sundaram, V.M.; Wen, S.B. Analysis of deep sub-micron resolution in microsphere based imaging. *Appl. Phys. Lett.* **2014**, *105*, 204102. [CrossRef]

79. Li, Y.; Liu, X.; Li, B. Single-cell biomagnifier for optical nanoscopes and nanotweezers. *Light Sci. Appl.* **2019**, *8*, 61. [CrossRef]
80. Minin, I.V.; Minin, O.V.; Pacheco-Peña, V.; Beruete, M. Subwavelength, standing-wave optical trap based on photonic jets. *Quantum Electron.* **2016**, *46*, 555. [CrossRef]
81. Minin, I.V.; Minin, O.V.; Cao, Y.; Liu, Z.; Geints, Y.E.; Karabchevsky, A. Optical vacuum cleaner by optomechanical manipulation of nanoparticles using nanostructured mesoscale dielectric cuboid. *Sci. Rep.* **2019**, *9*, 1–8. [CrossRef]
82. Bahng, J.H.; Jahani, S.; Montjoy, D.G.; Yao, T.; Kotov, N.; Marandi, A. Mie resonance engineering in meta-shell supraparticles for nanoscale nonlinear optics. *ACS Nano* **2020**, *14*, 17203–17212. [CrossRef]
83. Minin, I.V.; Minin, O.V.; Glinskiy, I.A.; Khabibullin, R.A.; Malureanu, R.; Lavrinenko, A.V.; Yakubovsky, D.I.; Arsenin, A.V.; Volkov, V.S.; Ponomarev, D.S. Plasmonic nanojet: An experimental demonstration. *Opt. Lett.* **2020**, *45*, 3244–3247. [CrossRef] [PubMed]
84. Xu, K.; Jiang, C.; Chen, P.; Wang, W.; Zhang, Z. Miniature self-mixing interferometer based on a microsphere superlens for microactuator displacement measurement. *Opt. Commun.* **2021**, *498*, 127212. [CrossRef]
85. Zhou, Y.; Hong, M. Formation of a three-dimensional bottle beam via an engineered microsphere. *Photonics Res.* **2021**, *9*, 1598–1606. [CrossRef]
86. Minin, I.V.; Geints, Y.E.; Zemlyanov, A.A.; Minin, O.V. Specular-reflection photonic nanojet: Physical basis and optical trapping application. *Opt. Express* **2020**, *28*, 22690–22704. [CrossRef]
87. Yue, L.; Yan, B.; Monks, J.N.; Dhama, R.; Wang, Z.; Minin, O.V.; Minin, I.V. A millimetre-wave cuboid solid immersion lens with intensity-enhanced amplitude mask apodization. *J. Infrared Milli. Terahz. Waves* **2018**, *39*, 546–552. [CrossRef]
88. Nguyen Pham, H.H.; Hisatake, S.; Minin, O.V.; Nagatsuma, T.; Minin, I.V. Enhancement of spatial resolution of terahertz imaging systems based on terajet generation by dielectric cube. *Appl. Photonics* **2017**, *2*, 056106. [CrossRef]
89. Minin, I.V.; Minin, O.V. Terahertz artificial dielectric cuboid lens on substrate for super-resolution images. *Opt. Quantum Electron.* **2017**, *49*, 326. [CrossRef]
90. Ang, A.S.; Karabchevsky, A.; Minin, I.V.; Minin, O.V.; Sukhov, S.V.; Shalin, A.S. 'Photonic Hook' based optomechanical nanoparticle manipulator. *Sci. Rep.* **2018**, *8*, 2029. [CrossRef] [PubMed]
91. Zhu, J.; Goddard, L.L. Spatial control of photonic nanojets. *Opt. Express* **2016**, *24*, 30444–30464. [CrossRef]
92. Gu, G.; Zhou, R.; Chen, Z.; Xu, H.; Cai, G.; Cai, Z.; Hong, M. Super-long photonic nanojet generated from liquid-filled hollow microcylinder. *Opt. Lett.* **2015**, *40*, 625–628. [CrossRef]
93. Minin, I.V.; Minin, O.V.; Geints, Y.E.; Panina, E.K.; Karabchevsky, A. Optical manipulation of micro-and nanoobjects based on structured mesoscale particles: A brief review. *Atmos. Oceanic Opt.* **2020**, *33*, 464–469. [CrossRef]
94. Cao, Y.; Liu, Z.; Minin, O.V.; Minin, I.V. Deep subwavelength-scale light focusing and confinement in nanohole-structured mesoscale dielectric spheres. *Nanomaterials* **2019**, *9*, 186. [CrossRef]
95. Minin, I.V.; Minin, O.V. Recent trends in optical manipulation inspired by mesoscale photonics and diffraction optics. *J. Biomed. Photonics Eng.* **2020**, *6*, 020301. [CrossRef]
96. Minin, O.V.; Minin, I.V.; Cao, Y. Optical magnet for nanoparticles manipulations based on optical vacuum cleaner concept. Saratov Fall Meeting 2020: Optical and Nanotechnologies for Biology and Medicine. *Int. Soc. Opt. Photonics* **2021**, *11845*, 118451G. [CrossRef]
97. Bekirov, A.R.; Luk'yanchuk, B.S.; Fedyanin, A.A. Virtual image within a transparent dielectric sphere. *JETP Lett.* **2020**, *112*, 341–345. [CrossRef]
98. Wang, Z.; Luk'yanchuk, B.; Yue, L.; Yan, B.; Monks, J.; Dhama, R.; Minin, O.V.; Minin, I.V.; Huang, S.; Fedyanin, A.A. High order Fano resonances and giant magnetic fields in dielectric microspheres. *Sci. Rep.* **2019**, *9*, 20293. [CrossRef] [PubMed]
99. Huckabay, H.A.; Wildgen, S.M.; Dunn, R.C. Label-free detection of ovarian cancer biomarkers using whispering gallery mode imaging. *Biosens. Bioelectron.* **2013**, *45*, 223–229. [CrossRef]
100. Gerard, D.; Wenger, J.; Devilez, A.; Gachet, D.; Stout, B.; Bonod, N.; Popov, E.; Rigneault, H. Strong electromagnetic confinement near dielectric microspheres to enhance single-molecule fluorescence. *Opt. Express* **2008**, *16*, 15297–15303. [CrossRef]
101. Aouani, H.; Schon, P.; Brasselet, S.; Rigneault, H.; Wenger, J. Two-photon fluorescence correlation spectroscopy with high count rates and low background using dielectric microspheres. *Biomed. Opt. Express* **2010**, *1*, 1075–1083. [CrossRef] [PubMed]
102. Yang, H.; Cornaglia, M.; Gijs, M.A. Photonic nanojet array for fast detection of single nanoparticles in a flow. *Nano Lett.* **2015**, *15*, 1730–1735. [CrossRef]
103. Lu, D.; Pedroni, M.; Labrador-Páez, L.; Marqués, M.I.; Jaque, D.; Haro-González, P. Nanojet trapping of a single sub-10 nm upconverting nanoparticle in the full liquid water temperature range. *Small* **2021**, *17*, 2006764. [CrossRef] [PubMed]
104. Xin, H.; Xu, R.; Li, B. Optical trapping, driving and arrangement of particles using a tapered fibre probe. *Sci. Rep.* **2012**, *2*, 818. [CrossRef] [PubMed]
105. Xin, H.; Li, Y.; Liu, X.; Li, B. Escherichia coli-based biophotonic waveguides. *Nano Lett.* **2013**, *13*, 3408–3413. [CrossRef] [PubMed]
106. Lin, L.; Kollipara, P.S.; Kotnala, A.; Jiang, T.; Liu, Y.; Peng, X.; Korgel, B.A.; Zheng, Y. Opto-thermoelectric pulling of light-absorbing particles. *Light Sci. Appl.* **2020**, *9*, 34. [CrossRef]
107. Xin, H.; Li, Y.; Xu, D.; Zhang, Y.; Chen, C.H.; Li, B. Single upconversion nanoparticle–bacterium cotrapping for single–bacterium labeling and analysis. *Small* **2017**, *13*, 1603418. [CrossRef] [PubMed]

108. Xin, H.; Li, B. Optical orientation and shifting of a single multiwalled carbon nanotube. *Light Sci. Appl.* **2014**, *3*, e205. [CrossRef]
109. Li, Y.C.; Xin, H.B.; Lei, H.X.; Liu, L.L.; Li, Y.Z.; Zhang, Y.; Li, B.J. Manipulation and detection of single nanoparticles and biomolecules by a photonic nanojet. *Light Sci. Appl.* **2016**, *5*, e16176. [CrossRef] [PubMed]
110. Gérard, D.; Devilez, A.; Aouani, H.; Stout, B.; Bonod, N.; Wenger, J.; Popov, E.; Rigneault, H. Efficient excitation and collection of single-molecule fluorescence close to a dielectric microsphere. *J. Opt. Soc. Am. B* **2009**, *26*, 1473–1478. [CrossRef]
111. Li, Y.; Xin, H.; Zhang, Y.; Lei, H.; Zhang, T.; Ye, H.; Saenz, J.J.; Qiu, C.W.; Li, B. Living nanospear for near-field optical probing. *ACS Nano* **2018**, *12*, 10703–10711. [CrossRef] [PubMed]
112. Xin, H.; Li, Y.; Li, B. Controllable patterning of different cells via optical assembly of 1D periodic cell structures. *Adv. Funct. Mater.* **2015**, *25*, 2816–2823. [CrossRef]
113. Li, Y.; Liu, X.; Xu, X.; Xin, H.; Zhang, Y.; Li, B. Red-blood-cell waveguide as a living biosensor and micromotor. *Adv. Funct. Mater.* **2019**, *29*, 1905568. [CrossRef]
114. Li, Y.; Liu, X.; Yang, X.; Lei, H.; Zhang, Y.; Li, B. Enhancing upconversion fluorescence with a natural bio-microlens. *ACS Nano* **2017**, *11*, 10672–10680. [CrossRef]
115. Chen, X.; Wu, T.; Gong, Z.; Li, Y.; Zhang, Y.; Li, B. Subwavelength imaging and detection using adjustable and movable droplet microlenses. *Photonics Res.* **2020**, *8*, 225–234. [CrossRef]
116. Devilez, A.; Bonod, N.; Stout, B. Near field dielectric microlenses. *Proc. SPIE* **2010**, *7717*, 771708. [CrossRef]
117. Yang, S.; Taflove, A.; Backman, V. Experimental confirmation at visible light wavelengths of the backscattering enhancement phenomenon of the photonic nanojet. *Opt. Express* **2011**, *19*, 7084–7093. [CrossRef]
118. Li, Y.; Xin, H.; Liu, X.; Zhang, Y.; Lei, H.; Li, B. Trapping and detection of nanoparticles and cells using a parallel photonic nanojet array. *ACS Nano* **2016**, *10*, 5800–5808. [CrossRef]
119. Yi, K.J.; Wang, H.; Lu, Y.F.; Yang, Z.Y. Enhanced Raman scattering by self-assembled silica spherical microparticles. *J. Appl. Phys.* **2007**, *101*, 063528. [CrossRef]
120. Dantham, V.R.; Bisht, P.B.; Namboodiri, C.K.R. Enhancement of Raman scattering by two orders of magnitude using photonic nanojet of a microsphere. *J. Appl. Phys.* **2011**, *109*, 103103. [CrossRef]
121. Xing, C.; Yan, Y.; Feng, C.; Xu, J.; Dong, P.; Guan, W.; Zeng, Y.; Zhao, Y.; Jiang, Y. Flexible Microsphere-embedded film for microsphere-enhanced Raman spectroscopy. *ACS Appl. Mater. Interfaces* **2017**, *9*, 32896–32906. [CrossRef]
122. Arya, A.; Laha, R.; Das, G.M.; Dantham, V.R. Enhancement of Raman scattering signal using photonic nanojet of portable and reusable single microstructures. *J. Raman Spectrosc.* **2018**, *49*, 897–902. [CrossRef]
123. Gašparić, V.; Ristić, D.; Gebavi, H.; Ivanda, M. Resolution and signal enhancement of Raman mapping by photonic nanojet of a microsphere. *Appl. Surf. Sci.* **2021**, *545*, 149036. [CrossRef]
124. Kang, D.; Pang, C.; Kim, S.M.; Cho, H.S.; Um, H.S.; Choi, Y.W.; Suh, K.Y. Shape-controllable microlens arrays via direct transfer of photocurable polymer droplets. *Adv. Mater.* **2012**, *24*, 1709–1715. [CrossRef]
125. Lee, J.Y.; Hong, B.; Kim, W.Y.; Min, S.K.; Kim, Y.; Jouravlev, M.V.; Bose, R.; Kim, K.S.; Hwang, I.C.; Kaufman, L.J.; et al. Near-field focusing and magnification through self-assembled nanoscale spherical lenses. *Nature* **2009**, *460*, 498–501. [CrossRef]
126. Murade, C.U.; Ende, V.D.; Mugele, F.G. High speed adaptive liquid microlens array. *Opt. Express* **2012**, *20*, 18180–18187. [CrossRef] [PubMed]
127. Bogucki, A.; Zinkiewicz, Ł.; Grzeszczyk, M.; Pacuski, W.; Nogajewski, K.; Kazimierczuk, T.; Rodek, A.; Suffczyński, J.; Watanabe, K.; Taniguchi, K.; et al. Ultra-long-working-distance spectroscopy of single nanostructures with aspherical solid immersion microlenses. *Light Sci. Appl.* **2020**, *9*, 48. [CrossRef] [PubMed]
128. Lee, S.; Li, L. Rapid super-resolution imaging of sub-surface nanostructures beyond diffraction limit by high refractive index microsphere optical nanoscopy. *Opt. Commun.* **2015**, *334*, 253–257. [CrossRef]
129. Fan, W.; Yan, B.; Wang, Z.; Wu, L. Three-dimensional all-dielectric metamaterial solid immersion lens for subwavelength imaging at visible frequencies. *Sci. Adv.* **2016**, *2*, e1600901. [CrossRef] [PubMed]
130. Dhama, R.; Yan, B.; Palego, C.; Wang, Z. Super-Resolution imaging by dielectric superlenses: TiO_2 metamaterial superlens versus $BaTiO_3$ superlens. *Photonics* **2021**, *8*, 222. [CrossRef]
131. Monks, J.N.; Yan, B.; Hawkins, N.; Vollrath, F.; Wang, Z. Spider silk: Mother nature's bio-superlens. *Nano Lett.* **2016**, *16*, 5842–5845. [CrossRef]
132. Li, J.; Liu, W.; Li, T.; Rozen, I.; Zhao, J.; Bahari, B.; Kante, B.; Wang, J. Swimming microrobot optical nanoscopy. *Nano Lett.* **2016**, *16*, 6604–6609. [CrossRef]
133. Zhou, J.; Lian, Z.; Zhou, C.; Bi, S.; Wang, Y. Scanning microsphere array optical microscope for efficient and large area super-resolution imaging. *J. Opt.* **2020**, *22*, 105602. [CrossRef]
134. Wen, Y.; Yu, H.; Zhao, W.; Li, P.; Wang, F.; Ge, Z.; Wang, X.; Liu, L.; Li, W.J. Scanning super-resolution imaging in enclosed environment by laser tweezer controlled superlens. *Biophys. J.* **2020**, *119*, 2451–2460. [CrossRef]
135. Krivitsky, L.A.; Wang, J.J.; Wang, Z.; Luk'yanchuk, B. Locomotion of microspheres for super-resolution imaging. *Sci. Rep.* **2013**, *3*, 3501. [CrossRef] [PubMed]
136. Wang, S.; Zhang, D.; Zhang, H.; Han, X.; Xu, R. Super-resolution optical microscopy based on scannable cantilever-combined microsphere. *Microsc. Res. Tech.* **2015**, *78*, 1128–1132. [CrossRef]

137. Allen, K.W.; Farahi, N.; Li, Y.; Limberopoulos, N.I.; Walker, D.E., Jr.; Urbas, A.M.; Liberman, V.; Astratov, V.N. Super-resolution microscopy by movable thin-films with embedded microspheres: Resolution analysis. *Ann. Phys.* **2015**, *527*, 513–522. [CrossRef]
138. Zhang, T.; Li, P.; Yu, H.; Wang, F.; Wang, X.; Yang, T.; Yang, W.; Li, W.J.; Wang, Y.; Liu, L. Fabrication of flexible microlens arrays for parallel super-resolution imaging. *Appl. Surf. Sci.* **2020**, *504*, 144375. [CrossRef]
139. Yan, Y.; Li, L.; Feng, C.; Guo, W.; Lee, S.; Hong, M. Microsphere-coupled scanning laser confocal nanoscope for sub-diffraction-limited imaging at 25 nm lateral resolution in the visible spectrum. *ACS Nano* **2014**, *8*, 1809–1816. [CrossRef] [PubMed]
140. Yang, H.; Moullan, N.; Auwerx, J.; Gijs, M.A. Super-resolution biological microscopy using virtual imaging by a microsphere nanoscope. *Small* **2014**, *10*, 1712–1718. [CrossRef]
141. Chernomyrdin, N.V.; Kucheryavenko, A.S.; Kolontaeva, G.S.; Katyba, G.M.; Karalkin, P.A.; Parfenov, V.A.; Gryadunova, A.A.; Norkin, N.E.; Smolyanskaya, O.A.; Minin, O.V.; et al. A potential of terahertz solid immersion microscopy for visualizing sub-wavelength-scale tissue spheroids. *Proc. SPIE* **2018**, *10677*, 106771Y. [CrossRef]
142. Wang, F.; Liu, L.; Yu, H.; Wen, Y.; Yu, P.; Liu, Z.; Wang, Y.; Li, W.J. Scanning superlens microscopy for non-invasive large field-of-view visible light nanoscale imaging. *Nat. Commun.* **2016**, *7*, 13748. [CrossRef] [PubMed]

Communication

Generation of Photonic Hooks from Patchy Microcylinders

Fen Tang [1,2,†], Qingqing Shang [3,†], Songlin Yang [4], Ting Wang [5], Sorin Melinte [6], Chao Zuo [7] and Ran Ye [1,2,7,*]

1   School of Computer and Electronic Information, Nanjing Normal University, Nanjing 210023, China; 201043034@njnu.edu.cn
2   School of Artificial Intelligence, Nanjing Normal University, Nanjing 210023, China
3   Key Laboratory for Opto-Electronic Technology of Jiangsu Province, Nanjing Normal University, Nanjing 210023, China; 201043036@njnu.edu.cn
4   Advanced Photonics Center, Southeast University, Nanjing 210096, China; yangsonglin@seu.edu.cn
5   College of Materials and Chemistry & Chemical Engineering, Chengdu University of Technology, Chengdu 610059, China; wangting@cdut.edu.cn
6   Institute of Information and Communication Technologies, Electronics and Applied Mathematics, Université Catholique de Louvain, 1348 Louvain-la-Neuve, Belgium; sorin.melinte@uclouvain.be
7   Smart Computational Imaging Laboratory (SCILab), School of Electronic and Optical Engineering, Nanjing University of Science and Technology, Nanjing 210094, China; zuochao@njust.edu.cn
*   Correspondence: ran.ye@njnu.edu.cn
†   These authors contributed equally to this work.

Abstract: The photonic hook (PH) is a new type of curved light beam, which has promising applications in various fields such as nanoparticle manipulation, super-resolution imaging, and so forth. Herein, we proposed a new approach of utilizing patchy microcylinders for the generation of PHs. Numerical simulation based on the finite-difference time-domain method was used to investigate the field distribution characteristics of the PHs. By rotating the patchy microcylinder, PHs with different curvatures can be effectively generated, and the PH with a bending angle of 28.4° and a full-width-half-maximum of 0.36 λ can be obtained from 1 μm-diameter patchy microcylinders.

Keywords: photonic hook; photonic jet; patchy particles; microspheres

Citation: Tang, F.; Shang, Q.; Yang, S.; Wang, T.; Melinte, S.; Zuo, C.; Ye, R. Generation of Photonic Hooks from Patchy Microcylinders. *Photonics* **2021**, *8*, 466. https://doi.org/10.3390/photonics8110466

Received: 20 September 2021
Accepted: 19 October 2021
Published: 22 October 2021

Publisher's Note: MDPI stays neutral with regard to jurisdictional claims in published maps and institutional affiliations.

1. Introduction

Photonic nanojets (PJs) are narrow, high-intensity, non-evanescent light beams generated at the shadow side of dielectric particles when irradiating the particles with light waves [1]. They can propagate a distance longer than the wavelength of the incident light while keeping a minimum beamwidth smaller than $\lambda/2$, λ is the wavelength of the incident light [2]. PJs show promising applications in various fields, such as micromaching [3], single-particle manipulation [4], optical sensing [5,6], super-resolution imaging [7–10], and so forth. Within this context, many efforts have been made to design dielectric particles that can generate PJs with special characteristics [11,12].

In 2015, Minin et al. theoretically discovered a new type of PJ, which has a curved structure similar to the shape of a hook [13]. They called the curved PJ as a photonic hook (PH). The PH is formed when electromagnetic waves pass through a dielectric trapezoid particle composed of a cuboid and a triangular prism. The combined effects of the phase velocity difference and the interference of the waves lead to a curved high-intensity focus [14]. The PH phenomenon was later experimentally observed in the terahertz and optical range [15,16]. It can also be applied to the generation of curved surface plasmons [17].

PHs can be effectively generated from dielectric particles with asymmetric geometries, such as dielectric trapezoids [14,15] and dielectric cylinders with glass cuboids inside [18], or by using particles with an asymmetric distribution of refractive index (RI), such as Janus particles [19], Janus microbar [20], and so forth. Rotating the particles or adjusting the shape of the particles can effectively change the characteristics of the generated PHs [19].

Special illumination conditions, such as partial illumination [16,21] and nonuniform illumination [22], can also be used to generate PHs. In this way, PHs can be generated using microcylinders with a symmetric geometry and a uniform RI distribution [16,22]. In addition to obtaining PHs in the transmission mode, Liu et al. proposed the formation of PHs in the reflection mode [23], in which they used dielectric-coated concave hemicylindrical mirrors to bend the reflected light beams. Geints et al. also proposed the formation of PHs in the specular-reflection mode under the oblique illumination of a super-contrast dielectric particle [24]. Moreover, multiple PHs can be effectively generated using twin-ellipse microcylinders [25], adjacent dielectric cylinders [26] and two coherent illuminations [27]. The PHs have promising applications in various fields, for example, nanoparticle manipulation and cell redistribution [12,28]. Recently, Shang et al. reported the super-resolution imaging using patchy microspheres [29]. Unlike conventional microspheres, which have a symmetric PJ, the patchy microspheres have a curved focusing and show an improved imaging performance due to the asymmetric illumination. Asymmetric illumination is a technique to enhance the imaging contrast in conventional bright-field microscopic systems [30], and now it is widely used in computational microscopic imaging to produce phase contrast [31]. In addition, Minin et al. reported the contrast-enhanced terahertz microscopy under the near-field oblique subwavelength illumination based on the PHs formed by dielectric mesoscale particles [32].

In this work, we show that the PHs can be generated using patchy particles of dielectric microcylinders that are partially covered with Ag thin films. Numerical simulation based on the finite-difference-time-domain (FDTD) method was performed to investigate the characteristics of the PHs. The spatial distribution of the Poynting vector and the streamlines of the energy flow in the simulated light field were given to illustrate the formation mechanism of the PHs. By adjusting the RI of the background, the diameter of the patchy microcylinder and the opening angle of the Ag films, PHs with various curvatures and intensity enhancement abilities can be effectively formed. In addition, the method of tuning PHs by rotating patchy microcylinders was also discussed in this paper.

2. Simulation Method

Figures 1a,b are the schematic drawing of the 3D stereogram and 2D sectional view of the investigated model. A dielectric microcylinder was created for two-dimensional simulation with the FDTD method using Lumerical FDTD Solutions. The top surface of the cylinder is covered with a 100 nm-thick Ag film. As shown in Figure 1b, an intense focusing of light will occur on the rear side of the cylinder when a P-polarized monochromatic plane wave ($\lambda = 550$ nm) propagating parallelly to the X axis passes through the cylinder. In this study, the RI of the cylinder is set to be 1.9, the same as the RI of $BaTiO_3$ (BTG), a high-index dielectric material widely used in microsphere-based applications [3,9]. The diameter of the cylinder varies between 1–35 μm and the RI of the background changes between 1.00–1.52. For the entire computational domain, non-uniform meshes with RI-dependent element size were used and all of them are smaller than $\lambda/50$. As shown in Figure 1b, the PH's degree of curvature is defined by the bending angle β, which is the angle between the two lines connecting the start point with the hot point, and the hot point with the end point of the PH, respectively. The hot point is defined as the point with the largest intensity enhancement (I_{max}), and the end point is defined as the point on the middle line of the PH with an intensity enhancement factor of I_{max}/e [16,23]. The hook height increment H and the subtense L of the curved photonic flux are also shown in Figure 1b.

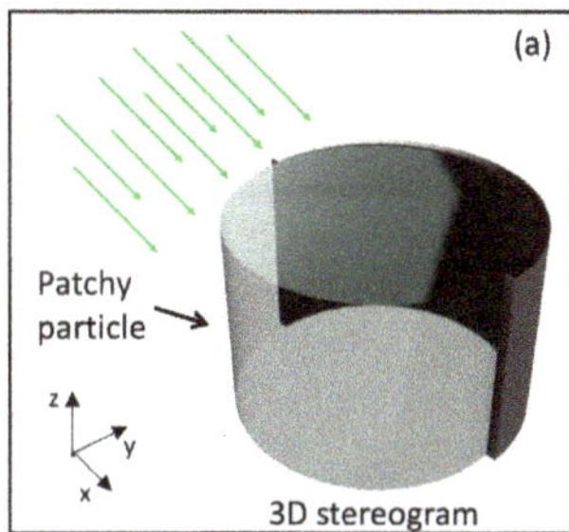
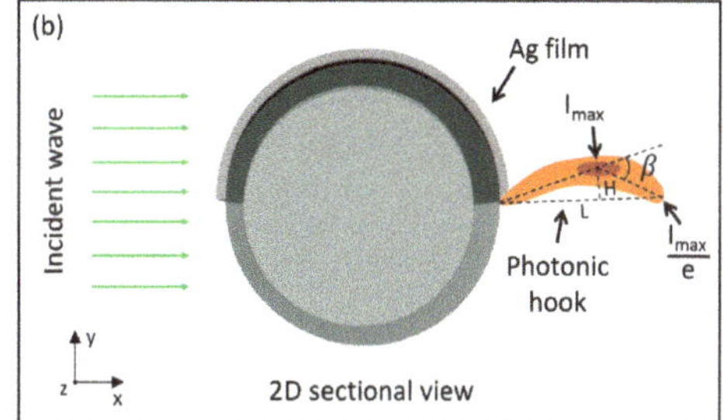

Figure 1. Schematics of a patchy microcylinder illuminated by plane waves: (**a**) 3D stereogram and (**b**) 2D sectional view.

3. Results and Discussion

First, we compared the optical field of the 35 μm-diameter pristine cylinder and patchy cylinder under plane wave illumination. The background medium here is microscope immersion oil (MIO, n_2 = 1.52). As shown in Figure 2a, the incident light passing through the pristine cylinder forms a conventional PJ on the shadow side of the cylinder. The generated PJ has a symmetric $|E|^2$ distribution with the the midline of the PJ as the center of symmetry. On the contrary, as for the patchy cylinder shown in Figure 2b, the upper part of the incident light is blocked by the Ag film covered on the top surface of the cylinder, so only the lower part of the incident light can enter the patchy cylinder. A curved light beam, that is, a PH, with a bending angle β = 12.5°, a hook height increment H = 1.51 μm and a subtense L = 28.81 μm is generated at the rear side of the cylinder. Compared with the pristine cylinder, the patchy cylinder has a smaller intensity enhancement ability, as the I_{max} is 36.0 and 14.3 for pristine and patchy cylinders, respectively.

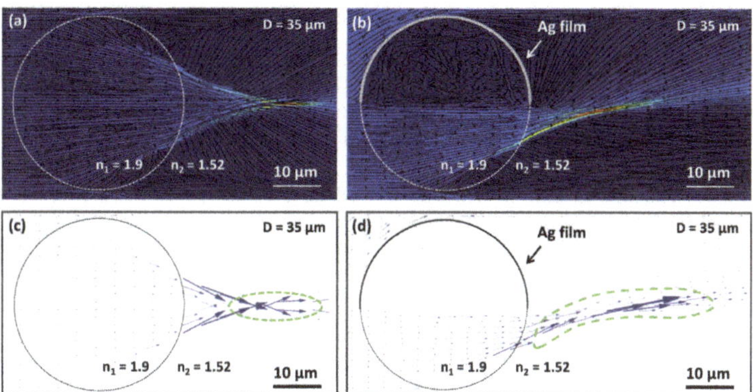

Figure 2. (**a**) PJ formed by a 35 μm-diameter pristine BTG microcylinder immersed in MIO; (**b**) PH formed by a 35 μm-diameter patchy BTG microcylinder immersed in MIO; (**c,d**) Corresponding Poynting vector of (**c**) the pristine microcylinder and (**d**) the patchy microcylinder.

As reported in the previous work [19,23], the formation mechanism of PJs and PHs can be analyzed using the time-averaged Poynting vector. In this work, the Poynting vector (blue conical arrows) of the optical field of the pristine and patchy cylinders under plane wave illumination is simulated with the FDTD method (Figure 2c,d), and the corresponding field-lines of the Poynting vector distribution are shown as the black lines in Figure 2a,b. As shown in Figure 2c, the spatial distribution of the Poynting vector inside and near the pristine cylinder is symmetric to the midline of the PJ (Figure 2c). Because the length of the conical arrows is proportional to the value of energy flux, the area containing longer arrows indicates a higher energy flux in that area. We can see that the energy flow corresponding to the pristine cylinder's optical field is focused into a classical PJ at the shadow side of the cylinder (Figure 2c). However, as for patchy cylinders, part of the incident light is reflected backwards to the space by the Ag film, which breaks the symmetry of illumination and makes the energy flow inside the microcylinder unbalanced. This asymmetric flow of energy is then focused into a curved beam after leaving the patchy cylinder, as shown in Figure 2d.

Next, the background medium was changed to air (n_2 = 1.0) (Figure 3a) and water (n_2 = 1.33) (Figure 3b) in order to investigate the influence of background RI on the characteristics of PHs. As shown in Figure 3a, when the background RI is 1.0, the light entering the microcylinder will be reflected multiple times by the Ag film. The direction of light propagation is thus changed from Path 1 to Path 2 on the first reflection, and then from Path 2 to Path 3 on the second reflection. When the background RI is 1.33, the patchy cylinder forms a jet-like beam with a bending angle β = 15.4°. The subtense and height increment of the beam are L = 6.17 μm and H = 0.30 μm, respectively. The beam has the greatest intensity outside the cylinder with an enhancement factor of I_{max} = 24.8. Considering that water is one of the most commonly used biocompatible materials, studying the PHs in water is of great importance for the practical applications of such curved beams. Therefore, we will use n_2 = 1.33 as the RI of the background medium in the following studies.

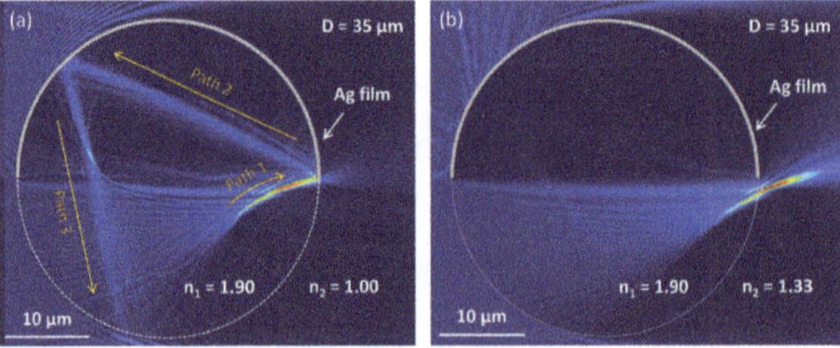

Figure 3. Optical fields of 35 μm-diameter patchy microcylinders immersed in (**a**) air and (**b**) water.

The influence of particle diameter on the characteristics of PHs is also investigated in this study. Patchy cylinders of various diameters between 1 μm and 20 μm are illuminated with plane waves (λ = 550 nm). The RI of the cylinder and the background medium is fixed at n_1 = 1.90 and n_2 = 1.33, respectively. As shown in Figure 4a, a 20 μm-diameter patchy cylinder generates a PH with a bending angle β = 17.6°, a hook height increment H = 0.52 μm and a subtense L = 7.19 μm. The cross-sectional intensity profiles of the PH retrieved from the orange dash lines also confirm the curved trajectory of the I_{max} position along the X axis (Figure 4b). We found that the PHs generated by smaller patchy cylinders have a slightly greater curvature and a smaller subtense length and height increment, as the 10 μm- and 5 μm-diameter patchy cylinders can produce PHs with β = 18.8°, L = 3.03 μm, H = 0.10 μm (Figure 4c) and β = 20.2°, L = 1.33 μm, H = 0.05 μm (Figure 4d), respectively. The intensity enhancement ability is found to be weaker in small particles. The I_{max} = 13.3 (Figure 4a), 10.4 (Figure 4c), 7.3 (Figure 4d) for 20 μm-, 10 μm-, 5 μm-diameter patchy cylinders, respectively. This difference in intensity enhancement ability is due to the fact that large particles can collect more light waves and focus them onto a narrow space, leading to a higher I_{max} at the focal point. However, there is no PH phenomenon in the optical field of patchy cylinders when the cylinder diameter is reduced to 2 μm (Figure 4e) and 1 μm (Figure 4f), because the light scattering of dielectric particles with a diameter close to the wavelength of the incident light tend to be localized in the forward direction and no jet-like fields can be generated [11].

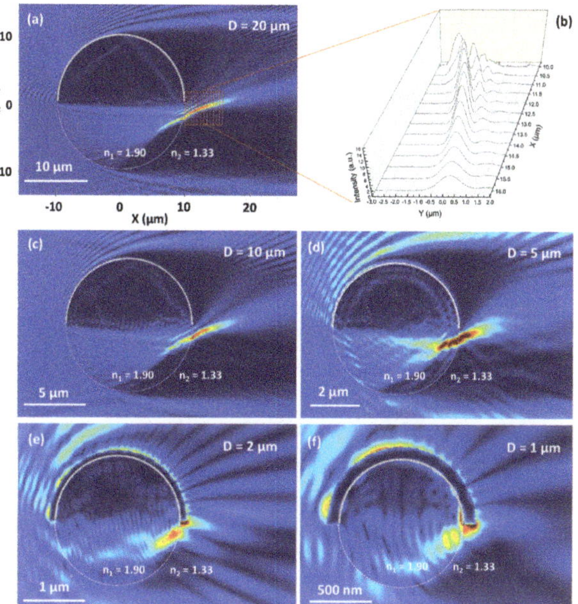

Figure 4. (**a**) Optical field of the 20 μm-diameter patchy microcylinder and (**b**) the cross-sectional profiles of the corresponding PH. (**c–f**) Optical fields of the patchy microcylinders with a diameter of (**c**) 10 μm, (**d**) 5 μm, (**e**) 2 μm and (**f**) 1 μm.

In this study, we showed that the curvature of PHs can be changed by rotating the patchy cylinder. As shown in Figure 5a, a 35 μm-diameter patchy cylinder (n_1 = 1.9) fully immersed in water (n_2 = 1.33) with 1/4 of its surface covered by Ag films (α = 90°) is used for demonstration. The patchy cylinder is rotated clockwise around the center of the cylinder. The rotation angle (θ) is defined as the angle between the horizontal line and the left edge of the Ag film. θ is negative when the left edge of the Ag film is lower than the horizontal line, and positive when the left edge of the Ag film is higher than the horizontal line. As shown in Figure 5a, the PH with a bending angle β = 12.1° is generated when the patchy cylinder has a rotation angle θ = −10°. Increasing θ from −10° to 30° leads to the formation of PHs with a higher curvature (Figure 5d). The PH with a maximum bending angle β = 23.4° can be generated at α = 30° (Figure 5b). Then, the curvature of the PHs becomes smaller as θ is further increased from 30° to 90° (Figure 5d). However, when θ is between 45° and 80°, the light beams formed at the shadow side of the patchy cylinder is similar to a PJ, and its intensity distribution is approximately symmetric to the midline of the model, as shown in the inset of Figure 5d. The bending angle of the PH decreases to β = 15.2° at θ = 90° (Figure 5c).

As shown by the black line in Figure 5e, the subtense L of the PH increases from 13.33 μm to 18.23 μm when θ increases from −10° to 20°, and then it decreases to 15.18 μm when θ further increases to 45°. When θ is between 45° and 80°, the focused light has a structure similar to a symmetric PJ (the inset of Figure 5d). The light beams show a curved shape again when θ is between 80° and 90° (Figure 5c), and the subtense L decreases from 17.68 μm to 15.10 μm when increasing θ from 80° to 90°. The PH's height increment H at different rotation angles θ is shown as the red dots in Figure 5e, and its changing trend is similar to that of the subtense L. As shown by the red line in Figure 5e, the maximum height increment is obtained at θ = 30° with H = 1.56 μm. Then, the height increment drops to H = 1.07 μm when θ increases from 30° to 45°. When θ is between 80° and 90°, the height increment decreases as the rotation angle increases, with a value of H = 1.32 μm (θ = 80°) and H = 0.97 μm (θ = 90°), respectively.

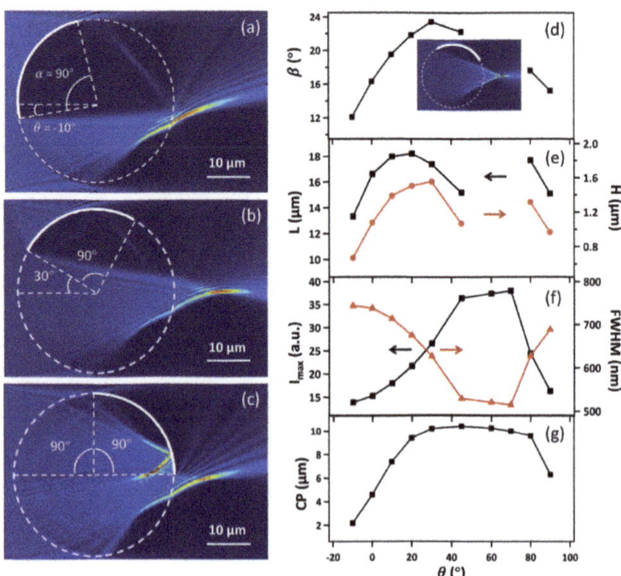

Figure 5. (**a–c**) PHs generated by the 35 µm-diameter patchy cylinder at different rotation angles: (**a**) $\theta = -10°$, (**b**) $\theta = 30°$, (**c**) $\theta = 90°$. (**d–g**) Characteristics of the PHs as a function of rotation angle θ: (**d**) bending angle β and the focus of a patchy particle at $\theta = 60°$ (the inset), (**e**) subtense L (black line) and hook height increment H (red line), (**f**) maximum intensity enhancement factor I_{max} (black line) and the corresponding FWHM (red line), (**g**) distance between the I_{max} position and the right edge of the cylinder (CP).

Figure 5f,g shows the value of I_{max} (black line in Figure 5f), the corresponding full width at half maximum (FWHM) (red line in Figure 5f) and the position of I_{max} (Figure 5g) at different rotation angles. The distance between the I_{max} position and the right edge of the cylinder is denoted CP. We found that the patchy cylinder has the greatest focusing ability (Figure 5f) and the farthest focal point (Figure 5g) when θ is between 45° and 70°, but the light beams focused by the patchy cylinder have a structure similar to a PJ, as shown in the inset of Figure 5d. Therefore, the patchy cylinder with a rotation angle $\theta = 30°$ shows the strongest bending ability as well as a good focusing performance.

Next, we changed the diameter of the patchy cylinders between 1 µm and 10 µm, while keeping the opening angle α of the Ag film as 90°, the rotation angle θ of the patchy cylinder as 30°, and the RI of the cylinder and the background medium as $n_1 = 1.90$ and $n_2 = 1.33$, respectively. As shown in Figure 6, in all four cases, the light beams with a curved structure are formed on the shadow side of the plane-wave illuminated patchy cylinder. Different from the results obtained using the patchy cylinders with half of their surfaces covered with Ag films ($\alpha = 180°$) (Figure 4e,f), the patchy cylinders with a smaller coverage of Ag film ($\alpha = 90°$) can generate PHs even when the diameter of the cylinder is below 5 µm (Figure 6c,d). The curvature β of the PH increases and the subtense L and the height increment H of the PH decrease as the diameter of the cylinder decreases. The characteristics of the PHs are $\beta = 21.2°$, L = 6.31 µm, H = 0.50 µm (Figure 6a), $\beta = 24.3°$, L = 3.26 µm, H = 0.21 µm (Figure 6b), $\beta = 26.9°$, L = 2.40 µm, H = 0.05 µm (Figure 6c) and $\beta = 28.4°$, L = 0.55 µm, H = 0.01 µm (Figure 6d) for 10 µm-, 5 µm-, 2 µm- and 1 µm-diameter patchy cylinders, respectively. We also found that the PHs generated from small particles have a smaller FWHM at the I_{max} position. In this study, the minimum FWHM at the I_{max} position is 196 nm, corresponding to $\sim 0.36\ \lambda$, which is generated from the 1 µm-diameter patchy cylinder (Figure 6d).

The patchy particles proposed in this work can be fabricated using the glancing angle deposition (GLAD) method. The GLAD is a technique that transports vapor deposition to a target at an inclined angle relative to the plane of the particle arrays [33,34]. In the GLAD process, particles within the same monolayer act as deposition masks for neighboring particles and thus patchy particles can be produced in a single deposition at a high yield of around 3.2×10^5 patchy particles per 1 cm^2 area for 2 μm-diameter microspheres. The GLAD technique allows the precise positioning of patches onto the particle surface by controlling geometric parameters like the deposition angle, the diameter of particles, and so forth. In addition, the GLAD technique can be applied to the deposition of functional patches with various optical properties, such as the anti-reflection coating, thin-film polarizers, and so forth. Multiple patches can be fabricated on a single particle using GLAD technology multiple times.

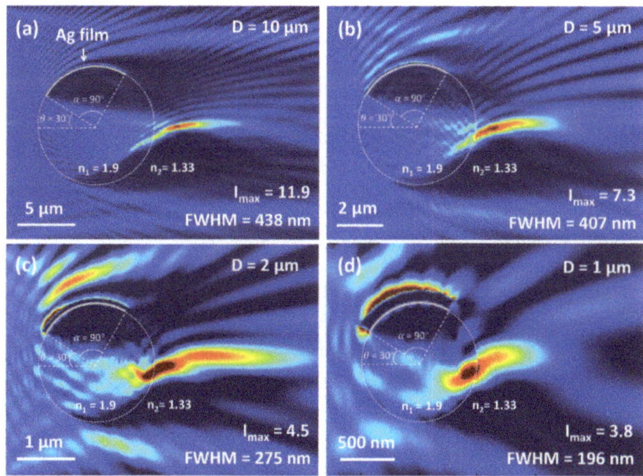

Figure 6. Optical fields of the patchy microcylinders with $\alpha = 90°$ and $\theta = 30°$. The diameter of the cylinder is (**a**) 10 μm, (**b**) 5 μm, (**c**) 2 μm and (**d**) 1 μm, respectively.

4. Conclusions

In conclusion, the generation of PHs from patchy microcylinders was investigated in detail in this study. The patchy microcylinders are dielectric cylinders whose surface is partially covered with Ag thin films. The fabrication of patchy cylinders can be realized using the GLAD method. Numerical simulation based on the FDTD method was used to investigate the characteristics of the PHs. By rotating a 35 μm-diameter patchy cylinder around its center, the bending angle of the PH can be changed between 12.1° and 23.7°, the subtense of the PH can be changed between 13.33 μm and 18.23 μm, and the height increment of the PH varies between 0.67 μm and 1.56 μm. PHs with a bending angle of 28.4° and a FWHM of 0.36 λ can be obtained by using a 1 μm-diameter patchy cylinder.

Author Contributions: Conceptualization, T.W., S.M. and R.Y.; methodology, F.T., Q.S. and R.Y.; software, F.T. and Q.S.; validation, Q.S. and S.Y.; formal analysis, F.T., Q.S. and R.Y.; writing—original draft preparation, R.Y.; supervision, C.Z. and R.Y.; funding acquisition, S.Y., F.T. and R.Y. All authors have read and agreed to the published version of the manuscript.

Funding: This research was funded by National Natural Science Foundation of China (62105156), Postgraduate Research & Practice Innovation Program of Jiangsu Province (SJCX21_0580, KYCX21_0102).

Institutional Review Board Statement: Not applicable.

Informed Consent Statement: Not applicable.

Data Availability Statement: The data presented in this study are available on reasonable request from the corresponding author.

Acknowledgments: The support of Fonds européen de développement régional (FEDER) and the Walloon region under the Operational Program "Wallonia-2020.EU" (project CLEARPOWER) is gratefully acknowledged.

Conflicts of Interest: The authors declare no conflict of interest.

References

1. Chen, Z.; Taflove, A.; Backman, V. Photonic nanojet enhancement of backscattering of light by nanoparticles: a potential novel visible-light ultramicroscopy technique. *Opt. Express* **2004**, *12*, 1214–1220. [CrossRef] [PubMed]
2. Heifetz, A.; Kong, S.C.; Sahakian, A.V.; Taflove, A.; Backman, V. Photonic Nanojets. *J. Comput. Theor. Nanosci.* **2009**, *6*, 1979–1992. [CrossRef]
3. Yan, B.; Yue, L.; Monks, J.N.; Yang, X.; Xiong, D.; Jiang, C.; Wang, Z. Superlensing plano-convex-microsphere (PCM) lens for direct laser nano-marking and beyond. *Opt. Lett.* **2020**, *45*, 1168–1171. [CrossRef]
4. Li, Y.C.; Xin, H.B.; Lei, H.X.; Liu, L.L.; Li, Y.Z.; Zhang, Y.; Li, B.J. Manipulation and detection of single nanoparticles and biomolecules by a photonic nanojet. *Light Sci. Appl.* **2016**, *5*, e16176. [CrossRef] [PubMed]
5. Yang, H.; Cornaglia, M.; Gijs, M.A.M. Photonic nanojet array for fast detection of single nanoparticles in a flow. *Nano Lett.* **2015**, *15*, 1730–1735. [CrossRef] [PubMed]
6. Gu, G.; Song, J.; Chen, M.; Peng, X.; Liang, H.; Qu, J. Single nanoparticle detection using a photonic nanojet. *Nanoscale* **2018**, *10*, 14182–14189. [CrossRef]
7. Wang, Z.; Guo, W.; Li, L.; Luk'yanchuk, B.; Khan, A.; Liu, Z.; Chen, Z.; Hong, M. Optical virtual imaging at 50 nm lateral resolution with a white-light nanoscope. *Nat. Commun.* **2011**, *2*, 218. [CrossRef]
8. Ye, R.; Ye, Y.H.; Ma, H.F.; Ma, J.; Wang, B.; Yao, J.; Liu, S.; Cao, L.; Xu, H.; Zhang, J.Y. Experimental far-field imaging properties of a ~5-μm diameter spherical lens. *Opt. Lett.* **2013**, *38*, 1829–1831. [CrossRef]
9. Wang, F.; Liu, L.; Yu, H.; Wen, Y.; Yu, P.; Liu, Z.; Wang, Y.; Li, W.J. Scanning superlens microscopy for non-invasive large field-of-view visible light nanoscale imaging. *Nat. Commun.* **2016**, *7*, 13748. [CrossRef]
10. Yang, H.; Trouillon, R.; Huszka, G.; Gijs, M.A.M. Super-resolution imaging of a dielectric microsphere is governed by the waist of its photonic nanojet. *Nano Lett.* **2016**, *16*, 4862–4870. [CrossRef]
11. Luk'Yanchuk, B.S.; Raniagua-Domínguez, R.; Minin, I.; Minin, O.; Wang, Z. Refractive index less than two: photonic nanojets yesterday, today and tomorrow. *Opt. Mater. Express* **2017**, *7*, 1820–1847. [CrossRef]
12. Minin, I.V.; Liu, C.Y.; Geints, Y.E.; Minin, O.V. Recent advances in integrated photonic jet-based photonics. *Photonics* **2020**, *7*, 41. [CrossRef]
13. Minin, I.; Minin, O. *Diffractive Optics and Nanophotonics: Resolution below the Diffraction Limit*; Springer: Berlin, Germany, 2015.
14. Yue, L.; Minin, O.V.; Wang, Z.; Monks, J.N.; Shalin, A.S.; Minin, I.V. Photonic hook: a new curved light beam. *Opt. Lett.* **2018**, *43*, 771–774. [CrossRef] [PubMed]
15. Minin, I.V.; Minin, O.V.; Katyba, G.M.; Chernomyrdin, N.V.; Kurlov, V.N.; Zaytsev, K.I.; Yue, L.; Wang, Z.; Christodoulides, D.N. Experimental observation of a photonic hook. *Appl. Phys. Lett.* **2019**, *114*, 031105. [CrossRef]
16. Minin, I.V.; Minin, Q.V.; Liu, C.Y.; Wei, H.D.; Geints, Y.E.; Karabchevsky, A. Experimental demonstration of a tunable photonic hook by a partially illuminated dielectric microcylinder. *Opt. Lett.* **2020**, *45*, 4899–4902. [CrossRef] [PubMed]
17. Minin, I.V.; Minin, O.V.; Ponomarev, D.S.; Glinskiy, I.A. Photonic hook plasmons: a new curved surface wave. *Ann. Phys.* **2018**, *530*, 1800359. [CrossRef]
18. Yang, J.; Twardowski, P.; Gérard, P.; Duo, Y.; Fontaine, J.; Lecler, S. Ultra-narrow photonic nanojets through a glass cuboid embedded in a dielectric cylinder. *Opt. Express* **2018**, *26*, 3723–3731. [CrossRef]
19. Gu, G.; Shao, L.; Song, J.; Qu, J.; Zheng, K.; Shen, X.; Peng, Z.; Hu, J.; Chen, X.; Chen, M.; Wu, Q. Photonic hooks from Janus microcylinders. *Opt. Express* **2019**, *27*, 37771–37780. [CrossRef]
20. Geints, Y.E.; Minin, I.V.; Minin, O.V. Tailoring 'photonic hook' from Janus dielectric microbar. *J. Opt.* **2020**, *22*, 065606. [CrossRef]
21. Liu, C.Y.; Chung, H.J.; Minin, O.V.; Minin, I.V. Shaping photonic hook via well-controlled illumination of finite-size graded-index micro-ellipsoid. *J. Opt.* **2020**, *22*, 085002. [CrossRef]
22. Gu, G.; Zhang, P.; Chen, S.; Zhang, Y.; Yang, H. Inflection point: a perspective on photonic nanojets. *Photonics Res.* **2021**, *9*, 1157–1171. [CrossRef]
23. Liu, C.Y.; Chung, H.J.; E, H.P. Reflective photonic hook achieved by a dielectric-coated concave hemicylindrical mirror. *J. Opt. Soc. Am. B Opt. Phys.* **2020**, *37*, 2528–2533. [CrossRef]
24. Geints, Y.E.; Zemlyanov, A.A.; Minin, I.V.; Minin, O.V. Specular-reflection photonic hook generation under oblique illumination of a super-contrast dielectric microparticle. *J. Opt.* **2021**, *23*, 045602. [CrossRef]
25. Shen, X.; Gu, G.; Shao, L.; Peng, Z.; Hu, J.; Bandyopadhyay, S.; Liu, Y.; Jiang, J.; Chen, M. Twin photonic hooks generated by twin-ellipse microcylinder. *IEEE Photonics J.* **2020**, *12*, 6500609. [CrossRef]
26. Zhou, S. Twin photonic hooks generated from two adjacent dielectric cylinders. *Opt. Quantum Electron.* **2020**, *52*, 389. [CrossRef]
27. Zhou, S. Twin photonic hooks generated from two coherent illuminations of a micro-cylinder. *J. Opt.* **2020**, *22*, 085602. [CrossRef]
28. Dholakia, K.; Bruce, G.D. Optical hooks. *Nat. Photonics* **2019**, *13*, 229–230. [CrossRef]
29. Shang, Q.; Tang, F.; Yu, L.; Oubaha, H.; Caina, D.; Melinte, S.; Zuo, C.; Wang, Z.; Ye, R. Super-resolution imaging with patchy microspheres. *arXiv* **2021**, arXiv:2108.06242.

30. Sanchez, C.; Cristóbal, G.; Bueno, G.; Blanco, S.; Borrego-Ramos, M.; Olenici, A.; Pedraza, A; Ruiz-Santaquiteria, J. Oblique illumination in microscopy: A quantitative evaluation. *Micron* **2018**, *105*, 47–54. [CrossRef] [PubMed]
31. Fan, Y.; Li, J.; Lu, L.; Sun, J.; Hu, Y.; Zhang, J.; Li, Z.; Shen, Q.; Wang, B.; Zhang, R.; Chen, Q.; Zuo, C. Smart computational light microscopes (SCLMs) of smart computational imaging laboratory (SCILab). *PhotoniX* **2021**, *2*, 19. [CrossRef]
32. Minin, O.V.; Minin, I.V. Terahertz microscopy with oblique subwavelength illumination in near field. *Preprints* **2021**, 2021090024. [CrossRef]
33. Pawar, A.B.; Kretzschmar, I. Patchy particles by glancing angle deposition. *Langmuir* **2008**, *24*, 355–358. [CrossRef] [PubMed]
34. Pawar, A.B.; Kretzschmar, I. Multifunctional patchy particles by glancing angle deposition. *Langmuir* **2009**, *25*, 9057–9063. [CrossRef] [PubMed]

*photonics*

Communication

Super-Resolution Imaging with Patchy Microspheres

Qingqing Shang [1,†], Fen Tang [2,3,†], Lingya Yu [4], Hamid Oubaha [5], Darwin Caina [5,6], Songlin Yang [7], Sorin Melinte [5], Chao Zuo [8], Zengbo Wang [4] and Ran Ye [2,3,8,*]

[1] Key Laboratory for Opto-Electronic Technology of Jiangsu Province, Nanjing Normal University, Nanjing 210023, China; 201043036@njnu.edu.cn
[2] School of Computer and Electronic Information, Nanjing Normal University, Nanjing 210023, China; 201043034@njnu.edu.cn
[3] School of Artificial Intelligence, Nanjing Normal University, Nanjing 210023, China
[4] School of Computer Science and Electronic Engineering, Bangor University, Bangor LL57 1UT, UK; yulingya0408@gmail.com (L.Y.); z.wang@bangor.ac.uk (Z.W.)
[5] Institute of Information and Communication Technologies, Electronics and Applied Mathematics, Université Catholique de Louvain, 1348 Louvain-la-Neuve, Belgium; hamid.oubaha@uclouvain.be (H.O.); darwin.caina@uclouvain.be (D.C.); sorin.melinte@uclouvain.be (S.M.)
[6] Faculty of Science, Universidad Central del Ecuador, Quito 170521, Ecuador
[7] Advanced Photonics Center, Southeast University, Nanjing 210096, China; yangsonglin@seu.edu.cn
[8] Smart Computational Imaging Laboratory (SCILab), School of Electronic and Optical Engineering, Nanjing University of Science and Technology, Nanjing 210094, China; zuochao@njust.edu.cn
*   Correspondence: ran.ye@njnu.edu.cn
†   These authors contributed equally to this work.

Abstract: The diffraction limit is a fundamental barrier in optical microscopy, which restricts the smallest resolvable feature size of a microscopic system. Microsphere-based microscopy has proven to be a promising tool for challenging the diffraction limit. Nevertheless, the microspheres have a low imaging contrast in air, which hinders the application of this technique. In this work, we demonstrate that this challenge can be effectively overcome by using partially Ag-plated microspheres. The deposited Ag film acts as an aperture stop that blocks a portion of the incident beam, forming a photonic hook and an oblique near-field illumination. Such a photonic hook significantly enhanced the imaging contrast of the system, as experimentally verified by imaging the Blu-ray disc surface and colloidal particle arrays.

Keywords: photonic jet; photonic hook; patchy particle; microspheres; super-resolution imaging

Citation: Shang, Q.; Tang, F.; Yu, L.; Oubaha, H.; Caina, D.; Yang, S.; Melinte, S.; Zuo, C.; Wang, Z.; Ye, R. Super-Resolution Imaging with Patchy Microspheres. *Photonics* **2021**, *8*, 513. https://doi.org/10.3390/photonics8110513

Received: 10 October 2021
Accepted: 8 November 2021
Published: 15 November 2021

Publisher's Note: MDPI stays neutral with regard to jurisdictional claims in published maps and institutional affiliations.

1. Introduction

Optical microscopes (OMs) are one of the most important tools for scientific research. Due to the Abbe diffraction limit, conventional OMs cannot resolve two objects closer than $0.5\lambda/NA$, where λ is the incident wavelength and NA is the numerical aperture of the microscope. Therefore, an OM equipped with a near-unity NA objective and a white light source ($\lambda \sim 550$–600 nm) has a resolution limit of 300 nm. Within this context, many different methods have been proposed to overcome this limitation. In 2009, Lee et al. successfully used nanoscale spherical lenses to resolve features beyond the diffraction limit [1]. Later, Wang et al. demonstrated the super-resolution imaging of 50 nm features using dielectric microspheres under white light illumination [2]. Smaller features, i.e., 15–25 nm nanogaps, can be resolved with microspheres under a confocal microscope [3,4]. Since the resolution of an imaging system is often measured by the point spread function (PSF) instead of by the resolved feature size, Allen et al. developed a convolution-based resolution analysis method and derived the best resolution for microsphere nanoscopy is $\sim\lambda/6$–$\lambda/7$ [4]. Microsphere-assisted imaging has the advantages of simple operation, no fluorescent labeling, and good compatibility with commercial OMs. To obtain high-quality images, various parameters affecting the imaging performance of microspheres have been studied, such as illumination conditions [5], microsphere diameters [6], immersion mode [7–9], and immersion materials [10–12].

Currently, most microsphere-assisted imaging methods use high-refractive-index microspheres in a liquid environment [13–17]. Nevertheless, the imaging performance of the microspheres can be affected by the shape of air–liquid interfaces as well as the refractive index distribution of liquid films. Moreover, samples may be contaminated or even damaged in liquid. Only a small amount of research has been done to improve the imaging performance of microspheres in air, such as improving illumination conditions [5], optimizing the diameter and the refractive index of microspheres [18], using plano-convex-microsphere (PCM) lens design [19,20], and using a microsphere lens group [21].

Patchy particles are anisotropic particles having two or more different physical or chemical properties on their surfaces. They are considered as promising materials for biomedical, environmentally friendly, and sustainable applications, such as biosensing, drug delivery, water decontamination, hydrogen production, etc. [22,23]. Patchy particles are also a type of optical functional materials. For example, the mirror made of Ag-coated patchy particles can have optical responses to external electric and magnetic fields, and its optical reflectivity can be adjusted in real-time by changing the external fields [24]. Here we present the performance of super-resolution imaging in air using patchy microspheres. To the best of our knowledge, this is the first study showing that patchy microspheres are suitable for super-resolution imaging. The patchy particles can generate a curved photonic jet, i.e., photonic hook, due to their structural asymmetry [25–27], which is shown to be useful in boosting the imaging contrast and quality in this work. The results will contribute to the further advancement of the microsphere-based optical nanoscopy/microscopy techniques and facilitate their applications in nanotechnology, life sciences, etc.

2. Materials and Methods

Figure 1a illustrates the schematic drawing of the patchy microsphere fabrication by glancing angle deposition (GLAD) method [28]. BaTiO$_3$ glass (BTG, 27–35 µm diameter, n = 1.9, Microspheres-Nanospheres, USA) were self-assembled into monolayers by drop-casting a small amount of BTG powders on a glass slide followed by using water to compact them together. The microsphere arrays were then coated with 100 nm thick Ag films by physical vapor deposition (PVD) (1 Å/s) at a constant angle ($\alpha = 60°$). We chose Ag because it has been used in microsphere-assisted microscopy to modify the optical properties of the substrate on which the sample is fixed [29] or the sample itself [30], in order to achieve the desired imaging quality. After deposition, a stream of deionized water was used to transfer the microspheres from the glass slide to the observation sample. We call the fabricated patchy BTG microspheres p-BTG particles.

The p-BTG particles were observed with a commercial reflected light microscope (Axio AX10, Carl Zeiss) for super-resolution imaging (Figure 1b). The microspheres collected near-field nanoscale information from a Blu-ray disk (BD) surface and generated a magnified real image above the sample, which was then captured by the objective lens. This is confirmed by the imaging plane that is above the substrate, in contrast to virtual imaging whose image plane is down into the substrate. The entire imaging process was performed in air. A 20 × objective (NA = 0.4, EC EPIPLAN, Carl Zeiss) was used for imaging with the p-BTG particles. The system was illuminated by a white light source (HAL 100, Carl Zeiss). All the experimental results were recorded using a high-speed scientific complementary metal-oxide-semiconductor (CMOS) camera (DFC295, Leica). The Leica Application Suite X (LAS X) software was used to take optical microscopic images with the camera and to measure the intensity profiles from the recorded images. The magnification factor (M) was obtained by dividing the period of the intensity profiles by the period of the BD pattern.

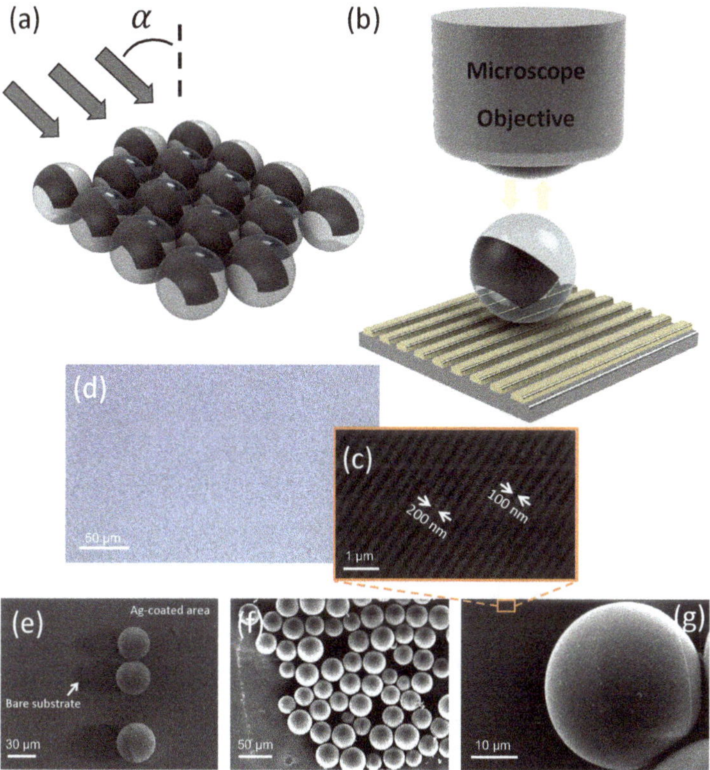

Figure 1. (**a**) Fabrication of patchy BTG (p-BTG) particle by glancing angle deposition method; (**b**) experimental setup of super-resolution imaging with p-BTG lens; (**c**) SEM image and (**d**) OM image of BD substrate (lines not resolved); (**e,f**) SEM images of p-BTG particle after Ag deposition; (**g**) SEM image of p-BTG particle transferred on BD substrate.

Figure 1c shows the scanning electron microscopic (SEM) image of the top surface of the BD sample in this study. It has a strip pattern with 300 nm periodicity, including 200 nm track width and 100 nm gap between two adjacent tracks. The pattern of the BD cannot be resolved by conventional OM method with diffraction-limited resolution of $\lambda/2\mathrm{NA} = 550/0.8 = 687.5$ nm (Figure 1d). The p-BTG microspheres after depositing 100 nm Ag by the GLAD method ($\alpha = 60°$) are shown in Figure 1e. There are some elliptical shadows on the left side of the microspheres (white arrow, Figure 1e), because the microspheres blocked the transportation of Ag vapor from the source to the substrate during deposition. The corresponding SEM image of the p-BTG particle arrays also confirms the presence of Ag patches on the microspheres (Figure 1f). Figure 1g shows a p-BTG on a BD sample, in which both the Ag patch and the strip pattern can be observed.

3. Results and Discussion

In this study, the imaging performance of BTG and p-BTG microlenses in air were compared with each other. As shown in Figure 2a, the BTG particle in air formed a magnified, low-contrast, real image of the strip pattern above the BD. The gap between the two neighboring tracks was 2.5 μm at the clearest image position, corresponding to a M of ~8.3×. Here, the super-resolution image formed by the BTG microsphere has a poor quality, with very low imaging contrast, which is not sufficient for most of the practical applications. On the contrary, as shown in Figure 2b, the new p-BTG microspheres generate significantly improved super-resolution images, both in quality and contrast. The imaging contrast has been boosted by a factor of ∼6.5, as shown in Figure 2c by retrieved intensity profiles along the white dash lines in Figure 2a,b. The measured magnification factor for the p-BTG lens is about 3.9×, which is smaller than that of unpatched particles (M = 8.3×). This is caused by different focusing characteristics of BTG and p-BTG particles which will be discussed below.

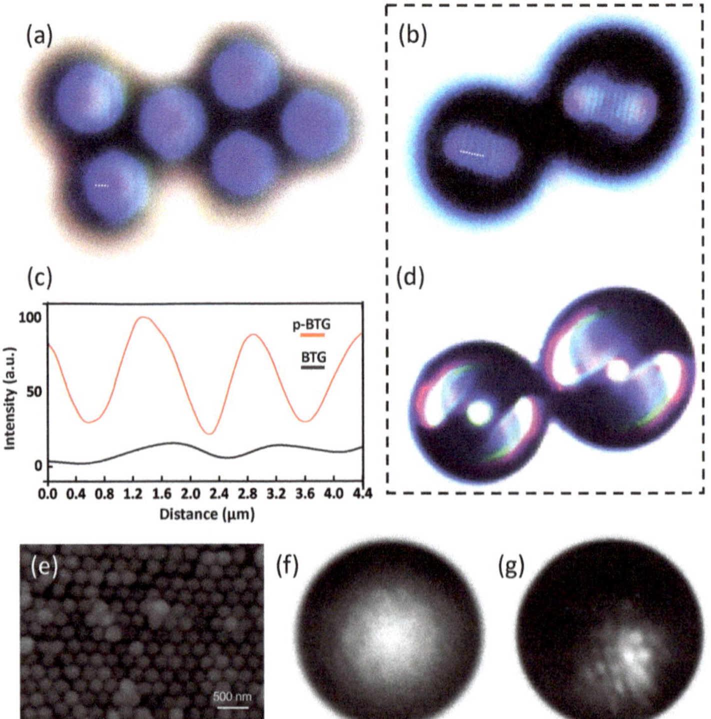

Figure 2. (**a**,**b**) OM images of the pattern on BD surface observed through (**a**) pristine BTG and (**b**) p-BTG with the 20 × objective; (**c**) optical intensity profiles across the white dash lines; (**d**) OM image of the corresponding p-BTG; (**e**) SEM image of the 230 nm-diameter SiO_2 particle arrays; (**f**,**g**) the SiO_2 particle arrays observed through (**f**) pristine BTG and (**g**) p-BTG microspheres.

Interestingly, as shown in Figure 2d, we observed two patchy textured patterns in each p-BTG particle, but only one side of the microspheres was coated with Ag films. The two paired patterns have a rotation angle of 180° between them around the center of the microsphere. This phenomenon could be attributed to the internal reflection occurring inside the high-index microspheres, in which case the Ag film prevents part of the incident light from entering the microsphere and causes a shadow with the same shape on the other side of the microsphere after multiple internal reflections.

In another imaging test, silica particles with 230 nm mean diameter (Nanorainbow Biotechnology, Nanjing, China) were self-assembled into arrays (Figure 2e) [31,32] and observed through BTG and p-BTG particles in air. The silica particle arrays were coated with 20 nm Ag before observation to enhance their reflectivity. As shown in Figure 2f,g, the p-BTG particle again demonstrates a better imaging performance over the BTG particle when imaging a sub-diffraction-limited nanoparticle array.

In our experiments, we found that the p-BTG particles deposited on the sample surface may have different appearances: bright, dark, or textured, and different imaging performance, depending on the positions of Ag coatings. As shown in Figure 3a, when the Ag film is at the bottom of the microsphere, it is like a concave mirror that reflects the incident light backward in a convergent way, so that more light can be collected by the objective, leading to a bright appearance. On the contrary, the p-BTG looks dark when the Ag film is on top of the microsphere. As shown in Figure 3b, the Ag film acts like a convex mirror that reflects the incident beam divergently at a large angle, so that most of the reflected light cannot be captured by the objective. The p-BTG lens shows a textured appearance and forms a magnified real image when the Ag film is on the side of the microsphere (Figure 3c), in which case the Ag film acts as an aperture stop, enhancing the contrast of the image and forming a photonic hook inside the microsphere. However, in this work the patchy particles have an uncontrolled movement during the water-flow-assisted transfer process, the position of the Ag coatings is random after the transfer step, so we took pictures of patchy particles with various appearances, and selected the most representative ones to be shown in Figure 3, in order to explain the observed optical phenomenon. Techniques such as probe-based microscale manipulation can be used to precisely control the positions of the Ag films in the future. In addition, we found that coating BTG particles with aluminum patches also can improve the microsphere's imaging contrast (Supplementary Materials Figure S1).

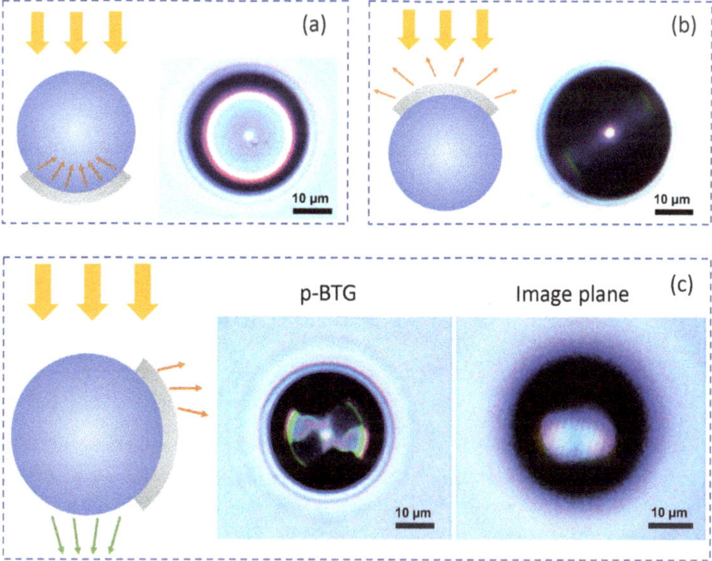

Figure 3. OM images of p-BTG particles when the Ag film is (**a**) at bottom of the microsphere, (**b**) on top of the microsphere, and (**c**) on the side of the microsphere.

To understand the main focusing properties of the p-BTG microsphere lens, computational modeling was performed on a workstation (HP Z8, 125 GB random access memory) using the two-dimensional (2D) finite-difference-time-domain (FDTD) method with Lumerical FDTD Solutions. 2D modeling is a commonly used method to investigate the light field around dielectric particles [25,27], because 3D sphere modeling is usually not possible due to the limited computing resource. As shown in Figure 4a,b, cylinders ($D = 35$ µm, $n_1 = 1.9$) were created for the FDTD simulation. The background medium was set to air ($n_2 = 1$). The area with the largest field intensity is considered as the focal point. The distance between the focal point and the bottom edge of the particle is defined as FP, which is positive when the focus is outside of the cylinder, and negative otherwise. As shown in Figure 4a, a plane light ($\lambda = 550$ nm) propagating in the Y direction forms a photonic jet inside the cylinder. Figure 4b is the FDTD simulation result of the intensity field distribution in the vicinity of a cylinder partially covered with a 100 nm thick Ag film ($\beta = 90°$, $\theta = 10°$). We can see the formation of a "photonic hook", with the light path off-centered and curved due to the asymmetry property of the incident beam caused by the Ag coating. In terms of imaging, the oblique illumination can help the lens to capture higher orders of diffraction from the sample [33] and enhance the phase contrast [34], which turned out to be very beneficial in boosting the quality and the contrast in microsphere-based super-resolution imaging. This enhancement mechanism can play an important role in developing more advanced and reliable microsphere super-resolution imaging systems.

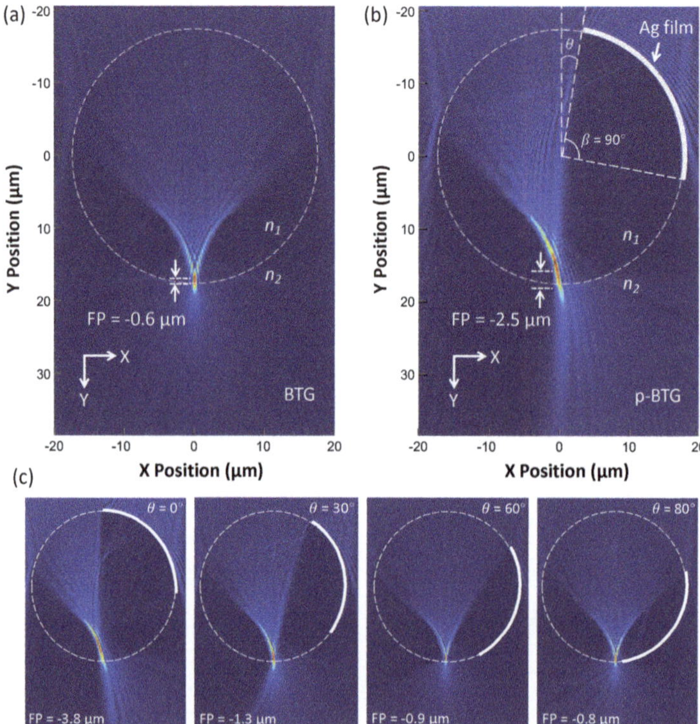

Figure 4. (**a,b**) FDTD-simulated light field of (**a**) the pristine BTG and (**b**) the p-BTG; (**c**) the influence of the position of Ag films on the focusing of the p-BTG.

From Figure 4b we can also see the angled "photonic hook" beam leads to a larger object-to-focus distance (O). Since magnification M = I/O, where I is the image plane position and O is the object plane position, increasing O will lead to decreased M which can explain why the p-BTG lens produces a smaller magnification factor as in experiments.

To illustrate the position effect of Ag films on BTG particle focusing, we varied the $θ$ angle in Figure 4b from 0° to 80°, while keeping the opening angle $β$ of the film coating as 90°. As shown in Figure 4c, the photonic hook phenomenon is maximized at $θ = 0$, which gradually decreases as $θ$ increases up to 60°. After this angle, the focusing does not show a curved hook focusing effect. Increasing $θ$ from 0° to 80° makes the focus move toward the bottom edge of the cylinder. When $θ$ is over 90°, the incident light is reflected back by the silver film and cannot reach the shadow side of the microsphere to illuminate the object (Figure S2). In our experiments, the p-BTG particle with $θ$ between 10° and 45° degrees is recommended for overall best performance which is a balanced choice between magnification factor and imaging contrast enhancement.

The focusing of the pristine and patchy particles was also studied with the ray-tracing method using the same model as for the FDTD simulation. A commercial software (TracePro, LAMBDA) was used to perform the simulation. As shown in Figure 5, the FP of the p-BTG has an approximately constant value of ~0.95 μm as the rotation angle θ of the Ag film increases from 0° to 80°. When θ is between 60° (Figure 5e) and 80° (Figure 5f), the p-BTG cylinders have a focusing performance similar to that of the pristine BTG (Figure 5a). Compared with the FDTD simulation, the ray tracing shows a longer focal length and the focus is outside of the particles. This difference has been discussed in the published literature [1,2], which reported that when particle size reduces to the super-resolution size window, the particle will have remarkably short near-field focal length [1], and geometrical ray tracing will become invalid and fail to predict the imaging properties for those super-resolution spheres, because light beams propagating through such small spheres could form optical vortices and singularities inside the sphere [2].

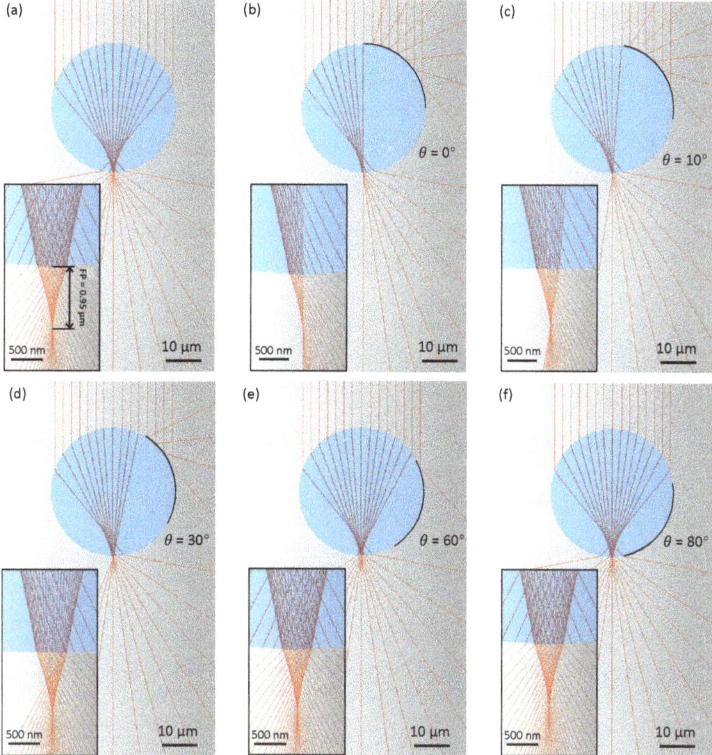

Figure 5. The focusing of pristine and patchy BTG particles in air simulated with the ray tracing method: (**a**) a pristine BTG; (**b–f**) p-BTG particles with Ag films at a rotation angle of (**b**) $\theta = 0°$, (**c**) $\theta = 10°$, (**d**) $\theta = 30°$, (**e**) $\theta = 60°$, and (**f**) $\theta = 80°$. The inset figures show the corresponding focal points.

In conclusion, BTG microspheres with patchy coating on their surface can provide a new strategy for improving the quality of super-resolution images obtained with high-index microspheres in air. Due to the formation of photonic hook illumination condition, the super-resolution imaging contrast can be improved by a factor of ~6.5, which significantly boosts the overall imaging quality. This method enables achieving high-quality super-resolution imaging without the use of immersion liquid, such as water or oil, opening a new path to developing more advanced and reliable nano-imaging systems based on engineered microsphere lenses. The method proposed in this work is in the early stages. To make a rational design of the decorated microlenses, we still need to understand the effects of the physical parameters (deposition material, film thickness, surface roughness, etc.) and the geometrical parameters (position, shape, area, etc.) of patches on microspheres' imaging performance. It is also important to develop compatible micromanipulators for the practical applications of the imaging system.

Supplementary Materials: The following are available at https://www.mdpi.com/article/10.3390/photonics8110513/s1, Figure S1: Optical microscopic image of a Blu-ray disc observed through an aluminum-coated BTG microsphere; Figure S2: The FDTD-simulated light field of a patchy cylinder with an opening angle β of 90° and a rotation angle θ of 120°.

Author Contributions: Conceptualization, R.Y.; methodology, F.T., Q.S., and R.Y.; software, L.Y. and F.T.; validation, Z.W., C.Z., and R.Y.; formal analysis, F.T., Q.S. and L.Y.; writing—original draft preparation, H.O., D.C., and R.Y.; supervision, C.Z. and R.Y.; funding acquisition, F.T., S.Y., S.M., C.Z., and R.Y. All authors have read and agreed to the published version of the manuscript.

Funding: This research was funded by National Natural Science Foundation of China (62105156), Postgraduate Research & Practice Innovation Program of Jiangsu Province (SJCX21_0580, KYCX21_0102), European Regional Development Fund (SPARCII-c81133).

Data Availability Statement: The data presented in this study are available on request from the corresponding author.

Acknowledgments: The support of Fonds Européen de développement régional (FEDER) and the Walloon region under the Operational Program "Wallonia-2020.EU" (project CLEARPOWER) is gratefully acknowledged.

Conflicts of Interest: The authors declare no conflict of interest.

References

1. Lee, J.Y.; Hong, B.H.; Kim, W.Y.; Min, S.K.; Kim, Y.; Jouravlev, M.V.; Bose, R.; Kim, K.S.; Hwang, I.C.; Kaufman, L.J.; et al. Near-field focusing and magnification through self-assembled nanoscale spherical lenses. *Nature* **2009**, *460*, 498–501. [CrossRef]
2. Wang, Z.; Guo, W.; Li, L.; Luk'Yanchuk, B.; Khan, A.; Liu, Z.; Chen, Z.; Hong, M. Optical virtual imaging at 50 nm lateral resolution with a white-light nanoscope. *Nat. Commun.* **2011**, *2*, 218. [CrossRef] [PubMed]
3. Yan, Y.; Li, L.; Feng, C.; Guo, W.; Lee, S.; Hong, M.H. Microsphere-coupled scanning laser confocal nanoscope for sub-diffraction-limited imaging at 25 nm lateral resolution in the visible spectrum. *ACS Nano* **2014**, *8*, 1809–1816. [CrossRef] [PubMed]
4. Allen, K.W.; Farahi, N.; Li, Y.; Limberopoulos, N.I.; Walker, D.E.; Urbas, A.M.; Liberman, V.; Astratov, V.N. Super-resolution microscopy by movable thin-films with embedded microspheres: Resolution analysis. *Ann. Phys.* **2015**, *527*, 513–522. [CrossRef]
5. Perrin, S.; Li, H.; Leong-Hoi, A.; Lecler, S.; Montgomery, P. Illumination conditions in microsphere-assisted microscopy. *J. Microsc.* **2019**, *274*, 69–75. [CrossRef]
6. Guo, M.; Ye, Y.H.; Hou, J.; Du, B. Size-dependent optical imaging properties of high-index immersed microsphere lens. *Appl. Phys. B-Lasers Opt.* **2016**, *122*, 15. [CrossRef]
7. Darafsheh, A.; Walsh, G.F.; Negro, L.D.; Astratov, V.N. Optical super-resolution by high-index liquid-immersed microspheres. *Appl. Phys. Lett.* **2012**, *101*, 141128. [CrossRef]
8. Wang, F.; Yang, S.; Ma, H.; Shen, P.; Wei, N.; Wang, M.; Xia, Y.; Deng, Y.; Ye, Y.H. Microsphere-assisted super-resolution imaging with enlarged numerical aperture by semi-immersion. *Appl. Phys. Lett.* **2018**, *112*, 023101. [CrossRef]
9. Zhou, Y.; Tang, Y.; He, Y.; Liu, X.; Hu, S. Effects of immersion depth on super-resolution properties of index-different microsphere-assisted nanoimaging. *Appl. Phys. Express* **2018**, *11*, 032501. [CrossRef]
10. Lee, S.; Li, L.; Wang, Z.; Guo, W.; Yan, Y.; Wang, T. Immersed transparent microsphere magnifying sub-diffraction-limited objects. *Appl. Opt.* **2013**, *52*, 7265–7270. [CrossRef]
11. Hao, X.; Kuang, C.; Liu, X.; Zhang, H.; Li, Y. Microsphere based microscope with optical super-resolution capability. *Appl. Phys. Lett.* **2011**, *99*, 10–13. [CrossRef]
12. Ye, R.; Ye, Y.-H.; Ma, H.F.; Cao, L.; Ma, J.; Wyrowski, F.; Shi, R.; Zhang, J.-Y. Experimental imaging properties of immersion microscale spherical lenses. *Sci. Rep.* **2014**, *4*, 3769. [CrossRef]
13. Ye, R.; Ye, Y.H.; Ma, H.F.; Ma, J.; Wang, B.; Yao, J.; Liu, S.; Cao, L.; Xu, H.; Zhang, J.Y. Experimental far-field imaging properties of a ~5-μm diameter spherical lens. *Opt. Lett.* **2013**, *38*, 1829–1831. [CrossRef]
14. Hou, B.; Xie, M.; He, R.; Ji, M.; Trummer, S.; Fink, R.H.; Zhang, L. Microsphere assisted super-resolution optical imaging of plasmonic interaction between gold nanoparticles. *Sci. Rep.* **2017**, *7*, 13789. [CrossRef] [PubMed]
15. Yang, H.; Moullan, N.; Auwerx, J.; Gijs, M.A. Super-resolution biological microscopy using virtual imaging by a microsphere nanoscope. *Small* **2014**, *10*, 1712–1718. [CrossRef] [PubMed]
16. Huszka, G.; Yang, H.; Gijs, M.A. Microsphere-based super-resolution scanning optical microscope. *Opt. Express* **2017**, *25*, 86–90. [CrossRef]

17. Li, L.; Guo, W.; Yan, Y.; Lee, S.; Wang, T. Label-free super-resolution imaging of adenoviruses by submerged microsphere optical nanoscopy. *Light. Sci. Appl.* **2013**, *2*, e104. [CrossRef]
18. Lee, S.; Li, L.; Ben-Aryeh, Y.; Wang, Z.; Guo, W. Overcoming the diffraction limit induced by microsphere optical nanoscopy. *J. Opt.* **2013**, *15*, 125710. [CrossRef]
19. Yan, B.; Yue, L.; Monks, J.N.; Yang, X.; Xiong, D.; Jiang, C.; Wang, Z. Superlensing plano-convex-microsphere (PCM) lens for direct laser nano-marking and beyond. *Opt. Lett.* **2020**, *45*, 1168–1171. [CrossRef]
20. Yan, B.; Song, Y.; Yang, X.; Xiong, D.; Wang, Z. Unibody microscope objective tipped with a microsphere: Design, fabrication, and application in subwavelength imaging. *Appl. Opt.* **2020**, *59*, 2641–2648. [CrossRef]
21. Luo, H.; Yu, H.; Wen, Y.; Zhang, T.; Li, P.; Wang, F.; Liu, L. Enhanced high-quality super-resolution imaging in air using microsphere lens groups. *Opt. Lett.* **2020**, *45*, 2981–2984. [CrossRef]
22. Su, H.; Price, C.-A.H.; Jing, L.; Tian, Q.; Liu, J.; Qian, K. Janus particles: Design, preparation, and biomedical applications. *Mater. Today Bio* **2019**, *4*, 100033. [CrossRef]
23. Marschelke, C.; Fery, A.; Synytska, A. Janus particles: From concepts to environmentally friendly materials ans sustainable applications. *Colloid Polym. Sci.* **2020**, *298*, 841–865. [CrossRef]
24. Aizawa, S.; Seto, K.; Tokunaga, E. External field response and applications of metal coated hemispherical Janus particles. *Appl. Sci.* **2018**, *8*, 653. [CrossRef]
25. Yue, L.; Minin, O.V.; Wang, Z.; Monks, J.N.; Shalin, A.S.; Minin, I.V. Photonic hook: A new curved light beam. *Opt. Lett.* **2018**, *43*, 771–774. [CrossRef]
26. Minin, I.V.; Minin, O.V.; Katyba, G.M.; Chernomyrdin, N.V.; Kurlov, V.N.; Zaytsev, K.I.; Yue, L.; Wang, Z.; Christodoulides, D.N. Experimental observation of a photonic hook. *Appl. Phys. Lett.* **2019**, *114*, 031105. [CrossRef]
27. Gu, G.; Shao, L.; Song, J.; Qu, J.; Zheng, K.; Shen, X.; Peng, Z.; Hu, J.; Chen, X.; Chen, M.; et al. Photonic hooks from Janus microcylinders. *Opt. Express* **2019**, *27*, 37771–37780. [CrossRef]
28. Pawar, A.B.; Kretzschmar, I. Patchy particles by glancing angle deposition. *Langmuir* **2008**, *24*, 355–358. [CrossRef]
29. Yang, S.; Cao, Y.; Shi, Q.; Wang, X.; Chen, T.; Wang, J.; Ye, Y.-H. Label-free super-resolution imaging of transparent dielectric objects assembled on a silver film by a microsphere-assisted microscope. *J. Phys. Chem. C* **2019**, *123*, 28353–28358. [CrossRef]
30. Cao, Y.; Yang, S.; Wang, J.; Shi, Q.; Ye, Y.-H. Surface plasmon enhancement for microsphere-assisted super-resolution imaging of metallodielectric nanostructures. *J. Appl. Phys.* **2020**, *127*, 233103. [CrossRef]
31. Ye, R.; Ye, Y.H.; Zhou, Z.; Xu, H. Gravity-assisted convective assembly of centimeter-sized uniform two-dimensional colloidal crystals. *Langmuir* **2013**, *29*, 1796–1801. [CrossRef] [PubMed]
32. Fang, C.; Yang, S.; Wang, X.; He, P.; Ye, R.; Ye, Y.H. Fabrication of two-dimensional silica colloidal crystals via a gravity-assisted confined self-assembly method. *Colloid Interface Sci. Commun.* **2020**, *37*, 100286. [CrossRef]
33. Sanchez, C.; Cristóbal, G.; Bueno, G.; Blanco, S.; Borrego-Ramos, M.; Olenici, A.; Pedraza, A.; Ruiz-Santaquiteria, J. Oblique illumination in microscopy: A quantitative evaluation. *Micron* **2018**, *105*, 47–54. [CrossRef] [PubMed]
34. Fan, Y.; Li, J.; Lu, L.; Sun, J.; Hu, Y.; Zhang, J.; Li, Z.; Shen, Q.; Wang, B.; Zhang, R.; et al. Smart computational light microscopes (SCLMs) of smart computational imaging laboratory (SCILab). *PhotoniX* **2021**, *2*, 19. [CrossRef]

 *photonics*

Article

Super-Resolution Imaging by Dielectric Superlenses: TiO_2 Metamaterial Superlens versus $BaTiO_3$ Superlens

Rakesh Dhama, Bing Yan, Cristiano Palego and Zengbo Wang *

School of Computer Science and Electronic Engineering, Bangor University, Bangor LL57 1UT, UK; dhama05@gmail.com (R.D.); b.yan@bangor.ac.uk (B.Y.); c.palego@bangor.ac.uk (C.P.)
* Correspondence: z.wang@bangor.ac.uk

Abstract: All-dielectric superlens made from micro and nano particles has emerged as a simple yet effective solution to label-free, super-resolution imaging. High-index $BaTiO_3$ Glass (BTG) microspheres are among the most widely used dielectric superlenses today but could potentially be replaced by a new class of TiO_2 metamaterial (meta-TiO_2) superlens made of TiO_2 nanoparticles. In this work, we designed and fabricated TiO_2 metamaterial superlens in full-sphere shape for the first time, which resembles BTG microsphere in terms of the physical shape, size, and effective refractive index. Super-resolution imaging performances were compared using the same sample, lighting, and imaging settings. The results show that TiO_2 meta-superlens performs consistently better over BTG superlens in terms of imaging contrast, clarity, field of view, and resolution, which was further supported by theoretical simulation. This opens new possibilities in developing more powerful, robust, and reliable super-resolution lens and imaging systems.

Keywords: super-resolution imaging; dielectric superlens; label-free imaging; titanium dioxide

Citation: Dhama, R.; Yan, B.; Palego, C.; Wang, Z. Super-Resolution Imaging by Dielectric Superlenses: TiO_2 Metamaterial Superlens versus $BaTiO_3$ Superlens. *Photonics* **2021**, *8*, 222. https://doi.org/10.3390/photonics8060222

Received: 12 May 2021
Accepted: 10 June 2021
Published: 15 June 2021

Publisher's Note: MDPI stays neutral with regard to jurisdictional claims in published maps and institutional affiliations.

1. Introduction

The optical microscope is the most common imaging tool known for its simple design, low cost, and great flexibility. However, the imaging resolution of a classical optical microscope is limited to almost half of its incident wavelength. The resolution limit in optics was discovered by the German physicist Ernst Abbe in 1873 by giving the expression, d = λ/(2NA) where d is the minimum distance between two structural elements as two objects instead of one, λ is the illuminating wavelength and NA is the numerical aperture of the used objective lens [1]. Such resolution limit is also known as Abbe's diffraction limit, which predicts the smallest objects that one can see through the objective lens of an optical microscope. The physical origin of diffraction limit arises due to the loss of exponentially decaying evanescent waves, which carry high spatial frequency subwavelength information from an object and are not able to propagate in the far-field [2]. This limits the resolution of subwavelength structures and results in an imperfect image. In this regard, several methods have been implemented to circumvent this resolution limit by converting these evanescent waves to propagating waves reaching the far-field. Near-field scanning optical microscope (NSOM) invented by D.W. Pohl is known as the first high-resolution imaging technique, which exploits nanoscale-sized tiny tip positioned close to specimen to collect the evanescent waves from the near-field and to transfer these lost subwavelength details into the far-field [3]. Such near-field imaging techniques require a longer time to acquire the image and cannot study the dynamic behavior of biological samples in real-time. Furthermore, experimental realization of Negative Index Medium (NIM) opened new opportunities towards super-resolution research when British scientist John Pendry theoretically showed how a slab of NIM can work as a perfect lens thanks to the enhancement of evanescent waves through the slab, instead of decaying [4–6]. Following this idea, several different types of plasmonic metamaterials lenses, such as super-lenses and hyper-lenses, have broken the diffraction limit [7–10]. However, these metal-based lenses have some serious limitations: (i) Exhibit high optical losses [11], (ii) involve complex and expensive nanofabrication process, (iii) involve the intense excitation of single visible wavelength laser and do not work under broadband white light sources. Besides, the development of super-resolution fluorescence optical microscope, which also won the 2014 Nobel prize in Chemistry, is another breakthrough to image biological cells and viruses beyond the diffraction limit [12]. This technique has also not been perfect due to its inability to resolve nonfluorescent samples, such as viruses and intracellular components, which cannot be labeled by fluorophores. In such a scenario, super-resolution through dielectric microspheres has emerged as a simple yet effective solution to all limitations mentioned above in other techniques. By clearly resolving sub-diffraction features in a plasmonic hexagonal nano array sample (50 nm holes separated 50 nm apart, with period 100 nm) through silica microspheres, we have reported microsphere nanoscopy working in real-time for the first time, which is label-free, usable under white light sources, and loss-free [13]. The technique was validated and resolution level of $\sim\lambda/6$–$\lambda/8$ was confirmed by other groups [14,15]. New microspheres, including Polystyrene and $BaTiO_3$ glass (BTG) microspheres (immersed in liquid or solid encapsulated), were soon introduced and widely used in the field [16,17]. The field has undergone rapid developments, including the development of scanning superlens, higher resolution metamaterial superlens [18,19], biological superlens [20], and integrated biochips, as well as new applications in interferometry, endoscopy, and others [21]. The underlying super-resolution mechanism is pretty complex and is still under investigation, which seems to be a mixing of photonic nanojet, optical super-resonances [22,23], illumination condition, and substrate effect, which were all summarized in our recent reviews [21,24,25]. Among these, it remains unclear which superlens will perform better, BTG or metamaterial superlens, which is a hemispherical all-dielectric lens made by 3D stacking of 15–20 nm titanium dioxide (TiO_2) nanoparticles ($n = 2.50$–2.55) following a bottom-up synthesis approach, with super-resolution of at least 45 nm have been reported under a white light microscope [18]. This paper will address it.

What we are interested in is designing the experiments to directly compare the super-resolution imaging performance between BTG superlens and TiO_2 superlens. Since many factors could influence the imaging process (e.g., substrate, illumination, particle size, shape, refractive index, etc.), we aim to design TiO_2 superlens to have the same effective refractive index as BTG particle, and perform comparative experiments under the same experimental conditions, including substrate, illumination, microscope, and imaging settings. This will allow a direct and reliable comparison of the imaging performance by both superlenses. Figure 1 shows the schematics of two microsphere superlens models, where BTG with refractive index of 1.92 was chosen, TiO_2 metamaterial (meta-TiO_2) superlens was designed and made from 20-nm sized TiO_2 ($n = 2.50$) via the modified bottom-up synthesis method described below, producing the same effective refractive index of 1.92 (corresponding to 61.3% TiO_2 volume ratio) as BTG sphere.

In this work, a novel, simpler, and repeatable method has been developed to fabricate proposed full-sphere meta-TiO_2 superlens as in Figure 1b. The sphere sizes can be controlled as the function of used air gun pressure during the fabrication procedure. These meta-TiO_2 microspheres exhibit good mechanical strength required for solid encapsulation with Polydimethylsiloxane (PDMS), which was applied to both BTG and TiO_2 superlens in our experiments before imaging. After PDMS encapsulation, a thin sheet with embedded superlenses was formed, which can be easily handled and moved at any desired location to image subwavelength nanofeatures.

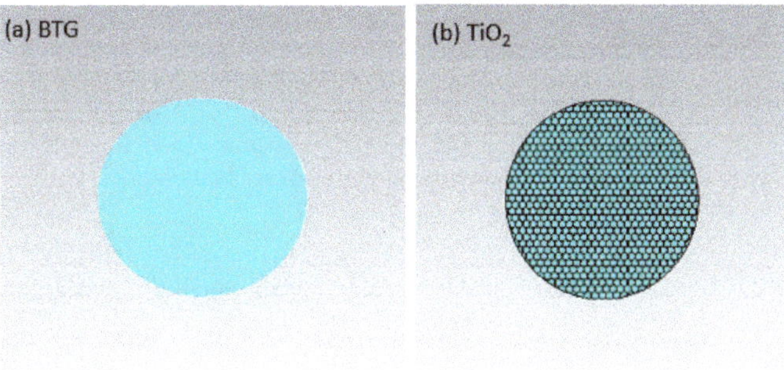

Figure 1. Schematic of BTG and TiO$_2$ microsphere models. (**a**) BTG microsphere with n = 1.92, and (**b**) meta-TiO$_2$: TiO$_2$ metamaterial microsphere, made from densely packed 20-nm, n = 2.50, TiO$_2$ nanoparticles, having the same effective refractive index of 1.92, as BTG particle.

2. Materials and Methods

In a previous study, we have shown that 15–20 nm TiO$_2$ nanoparticles should be used as the building block of the meta-superlens for best performance. Larger nanoparticles reduce the superlens performance [18]. Our metamaterial superlens fabrication method is based on the bottom-up approach of TiO$_2$ nanoparticles (20 nm, refractive index n = 2.50) to assemble into densely packed structures. The step-by-step procedure of fabricating these microspheres has been illustrated in Figure 2a, which is as follows. First, an aqueous solution of 20 nm anatase TiO$_2$ nanoparticles (from XuanChengJingRui New Material Co. Ltd., Anhui, China) is centrifuged at 17,500 rpm for 20 min at 10 °C to remove present aggregates and uneven size of nanoparticles. The unprecipitated solution containing a similar size of nanoparticles is centrifuged for the second time at 24,500 rpm for 60 min at 10 °C to obtain a densely packed precipitate. Then, the resulting supernatant is immediately removed from the precipitate to prevent the closed packed nature of nanoparticles. Then, this TiO$_2$ precipitate is diluted with deionized water by the weight ratio of 2:1 and left for 3 h to form TiO$_2$ gel. Meanwhile, a water-immiscible organic solvent mixture consisting of hexane and tetrachloroethylene with a volume ratio of 1:2 is used to cover the surface of a glass petri dish. Lastly, the TiO$_2$ gel is loaded into the container of an air gun and sprayed with an angle of 45° on the organic mixture layer spread on the petri dish surface. The organic mixture layer enables the densely packed nanoparticles to float on the oil/water interface, which undergoes a phase transition resulting in spherical structures after evaporation [18,26]. Figure 2b–d demonstrates how the increase in applied pressure (0.5 bar, 1 bar, and 1.5 bar) of air gun results in smaller-sized microspheres.

The fabricated meta-TiO$_2$ microspheres are sprayed on a silicon chip and left to evaporate the water for 1 h. Later, BTG microspheres are spread on the same chip and 10 nm thick gold layer is coated on nanochip sample containing TiO$_2$ and BTG microspheres together. A scanning electron microscope (SEM, Hitachi TM4000) is employed to confirm the spherical shapes of TiO$_2$ microspheres and to compare with BTG ones.

For super-resolution imaging, both BTG and meta-TiO$_2$ microsphere are immersed in a transparent host material to create a proper super-resolution imaging condition, which generates magnified virtual images of underlying nano-objects [21,27,28]. In this study, the PDMS host was used. To do this, Sylgard silicone elastomer and its curing agent was mixed in the weight ratio of 10:1 and stirred for 10 min. Next, this solution was placed in a vacuum chamber to remove bubbles. Meanwhile, meta-TiO$_2$ particles were fabricated on a petri dish using 1 bar air gun pressure as described above and 15–20 μm sized BTG microsphere was spread on another dish. Then, the same weight of PDMS solution was poured on both dishes and left out overnight (without heating to avoid potential influence on optical properties) to solidify this polymeric solution, which resulted in the formation of two 400 μm thick PDMS films containing meta-TiO$_2$ and BTG microspheres.

The fabricated film with embedded BTG and meta-TiO$_2$ microspheres was then placed on top of a semiconductor wafer containing different sizes and spacings of nanopatterns. Such samples were examined in reflection mode under a low-cost white light microscope (ICM100) using a 50× objective lens (NA = 0.6). In this regard, two nanopatterns of the semiconductor wafer and the comparison in their super-resolution capability through BTG and meta-TiO$_2$ microspheres as described below.

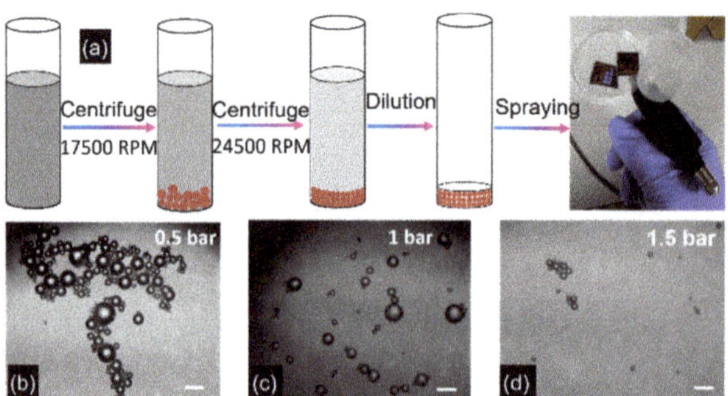

Figure 2. (**a**) Schematic illustration of fabrication procedure of TiO$_2$ microspheres, starting from two separate centrifuges, followed by dilution and spraying using an air gun. (**b–d**) Fabricated TiO$_2$ microspheres as the function of applied air gun pressure using 0.5, 1, and 1.5 bar, respectively. Scale bar: 20 μm.

3. Results

3.1. SEM Images of Meta-TiO$_2$ and BTG Microsphere Superlens

Figure 3 shows SEM image of fabricated meta-TiO$_2$ particles versus BTG particles on the same sample. The spheres can be easily distinguished based on the two simple observations as in Figure 3a: (1) Self-assembled TiO$_2$ microspheres have large variation in their sizes with respect to BTG spheres and (2) BTG microspheres appear brighter due to the larger scattering of secondary electrons towards detector in comparison to TiO$_2$ spheres. Furthermore, another nanochip sample of TiO$_2$ microspheres has been mounted on a 60° tilted holder to see the side view of the fabricated microspheres. In this context, the combined image Figure 3b confirms the full spherical shapes of TiO$_2$ microspheres sprayed on different nanofeatures of nanochip, this is the first report of such full-sphere TiO$_2$ particle superlens made by the fabrication approach described above, different from previously reported hemisphere TiO$_2$ superlens [18].

3.2. Super-Resolution Imaging Comparison

We now proceed to super-resolution imaging comparative results. Figure 4a shows the SEM image of meta-TiO$_2$ lens located on wafer pattern of 400 nm-sized nano-discs with a lattice spacing of 225 nm, which is more clearly shown in the inset of the figure. This pattern can be resolved by both BTG and meta-TiO$_2$ lens (size ~20–25 μm) as shown in Figure 4b,c, respectively, which demonstrates that meta-TiO$_2$ microspheres resolve the pattern with sharp contrast in comparison to BTG microspheres (for maximized imaging contrast, light illumination was tilted by 15-deg from a normal incident in Figure 4b,c,e,f). Furthermore, the same experiment has been performed on another pattern with the smaller features of 109 nm, as shown in Figure 4d. Comparing Figure 4e,f, we can clearly see that meta-TiO$_2$ lens surpasses BTG lens in imaging contrast, clarity, and field of view.

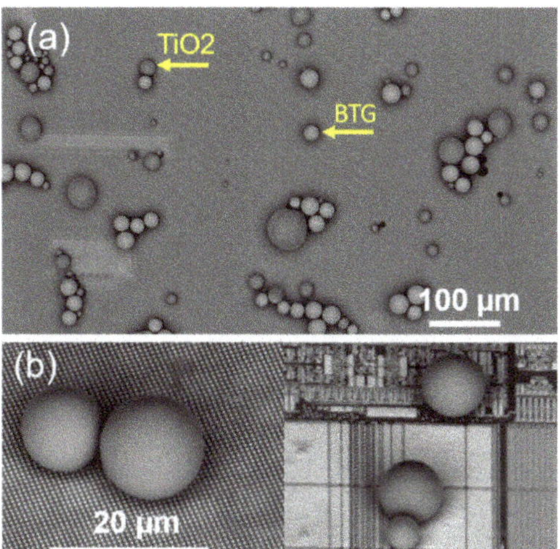

Figure 3. (a) SEM image of meta-TiO$_2$ (grey color) and BTG microspheres (bright) as indicated by arrows, separately. (b) A combined image of meta-TiO$_2$ microspheres located on different nanofeatures to demonstrate their full-sphere shapes.

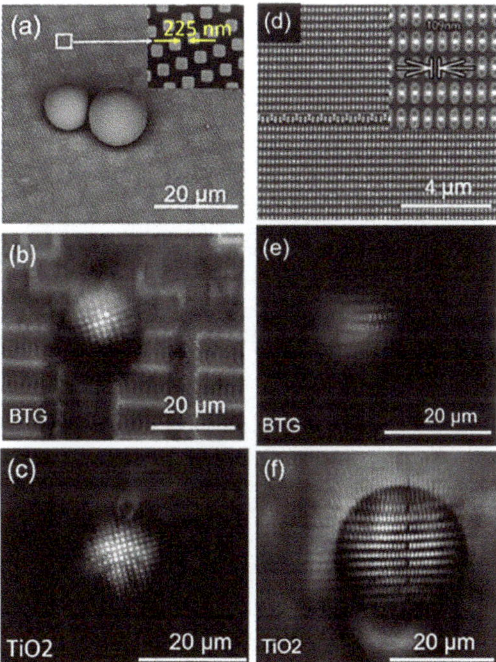

Figure 4. (a) SEM image of Meta-TiO$_2$ microspheres located on a wafer pattern of 400 nm features with a lattice spacing of 225 nm (Inset: Magnified image of the marked region). (b,c) Optical microscopic images of wafer pattern shown in (a) through BTG and TiO$_2$ microspheres, respectively. (d) SEM image of wafer pattern with 109 nm spacings, while the inset shows the magnified image of the same nanopattern. (e,f) Optical microscopic images of wafer pattern shown in (d) through BTG and TiO$_2$ microspheres, respectively.

For better comparison on imaging resolution, we have further carried out our imaging experiments using 90 nm line features on a wafer with a lattice spacing of 90 nm, which lies well below diffraction limit (~458 nm grating resolution, 229 nm feature resolution, for NA = 0.6 at 550 nm white-light peak wavelength), as shown in Figure 5a and its inset. Such 90 nm line features can hardly be resolved through the BTG microsphere, as presented in Figure 5b, while the super-resolved images of these 90-nm spaced line nanopatterns can be obtained through meta-TiO$_2$ microspheres, as shown in Figure 5c.

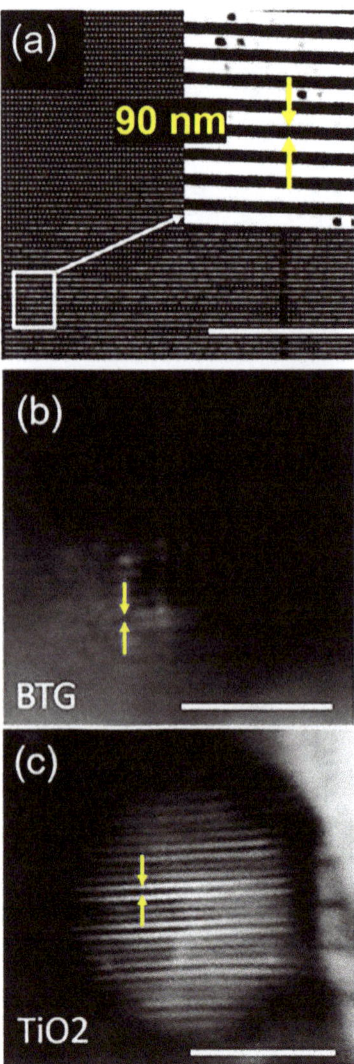

Figure 5. (**a**) SEM image of wafer pattern with 90 nm spacings, while the inset shows the magnified image of marked region. (**b,c**) Optical microscopic images of wafer pattern shown in (**a**) through BTG and TiO$_2$ microspheres, respectively. Scale bar: 10 μm.

4. Theory and Discussion

At nanoscale, the synthesized meta-TiO$_2$ lens surface is not smooth and non-homogenous, consisting of nanoparticles and air voids between them (Figure 1b). Our previous research has shown such composite lens has a particle volume fraction of 61.3% (close to theoretical close-packing limit of ~64%), with an effective refractive index of 1.92 (for n = 2.50)–1.95 (for n = 2.55) [18]. In comparison, the BTG lens is formed by smooth and homogenous material, with refractive index 1.92 chosen to match the index of the fabricated TiO_2 lens. As a result, for similar-sized TiO_2 and BTG lenses, their magnification factors will be close to each other. This was confirmed in experiments, as shown in Figure 4. The better imaging performance of TiO_2 lens over BTG lens thus cannot be satisfactorily explained by the classical effective medium theory (EMT), as both have the same effective index of 1.92. Instead, the fundamentals must be related to the non-homogeneous nature in nanoscale of the TiO_2 composite medium. It is virtually impossible to build an exact 3D model for the TiO_2 lens, which consists of millions of nanoparticles. The model needs to be simplified to a level our computer can handle. We simply 3D meta-sphere as a 2D cylinder model, with the top of the cylinder further simplified as a homogenous medium using EMT theory, while the bottom part of cylinder remains as nanoparticle stack. This simplification was illustrated in Figure 6b. The model was then built and simulated by full-wave software CST Microwave Studio. By placing an electric dipole source (z-polarized, perpendicular to the plane) close to the bottom of the particle lenses, we can analyze how the particle superlenses (for both BTG and TiO_2) collect dipole radiation energy (including evanescent and propagating wave components) and direct them into the far-field. To ensure a direct comparison is possible, all modeling settings were kept the same for both particle lens cases. The E-field distribution in Figure 6c,d shows both particles work similarly in directing the radiation energy into the far-field, which is also evidenced from far-field pattern in Figure 6i. However, we can see TiO_2 lens collects a bit more energy than BTG lens, as evidenced by the longer main lobe in far-field pattern in Figure 6i. This difference comes from the root of near-field interaction between dipole source and particle lens. From Figure 6e,f, we can see the root of the difference in near-field scattering, the TiO_2 nanoparticles in composite medium generate photonic nanojets array with jet dimension similar to a particle size of 20 nm (see Figure 6f,g), such high-frequency modulation provides a channel for evanescent wave components to be carried on top of propagating wave components into the far-field, which is more clearly seen in extracted data in Figure 6h. When the dipole source is x-polarized, the far-fields (Figure 6i) tend to be reduced while near-field jet strength tends to be increased (Figure 6g). However, for a 3D sphere model, both z-polarized and x-polarized dipoles will produce the same near-field and far-field results due to symmetry in a spherical object. In theory, the ultimate resolution (defined as feature resolution) of TiO_2 meta-lens will be 20 nm, equal to its composition particle size of 20 nm in the current case, but in reality, manufacturing imperfection and other factors, such as non-ideal contact between imaging objects and lens, will reduce the practical resolution, in our current case, with a full-sphere TiO_2 lens, the feature resolution is about 90 nm (grating resolution 180 nm). Better feature resolution of 45 nm has previously been achieved for hemisphere TiO_2 lens owing to much better contact between lens and imaging objects [18]. Overall, it is reasonable to conclude that nanocomposite medium design is beneficial for super-resolution development and represents a promising solution to the future development of more powerful, robust, and reliable dielectric superlens. The disadvantage is that it will be limited to surface near-field imaging and difficult to expand as 3D super-resolution imaging. In the next step, we are looking to incorporate upconversion nanoparticles (UCNPs) into the synthesis process of meta-TiO_2 superlens for biomedical super-resolution imaging applications.

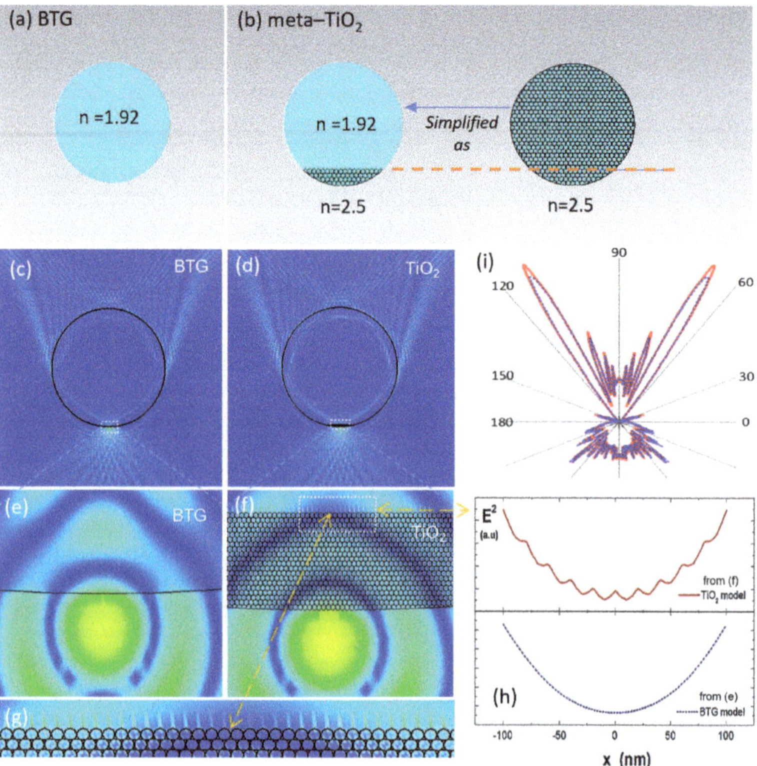

Figure 6. Modeling and comparison of electric field distribution of 20-μm-diameter (**a**) BTG and (**b**) meta-TiO$_2$ micro lenses under electric dipole source excitation. Field distribution for dipole source scattered by (**c**) BTG particle, and (**d**) TiO$_2$ particle. (**e**) Magnified view of near-field distribution in (**c**) marked zone. (**f**) Magnified view of near field around dipole source in (**d**) marked area. Noting the formation of tiny photonic jet array in (**f**), and (**g**) corresponding magnified view. (**h**) 1D plots of near-field intensity along indicated location in (**f**), and the same location in (**e**) (not shown). (**i**) Far-field scattering diagram for BTG (blue dash) and TiO$_2$ (red solid), showing the same radiation pattern for both lenses but with large intensity at the main lobe for TiO$_2$.

5. Conclusions

In summary, this work has confirmed the superiority of using TiO$_2$ metamaterial lens in super-resolution imaging development over BTG microsphere lens, with enhanced imaging contrast, clarity, and resolution under similar experimental conditions. It provides a solid foundation to the development of next-generation, more powerful, robust, and reliable optical nano-imaging platforms based on dielectric super lenses.

Author Contributions: Conceptualization, Z.W. and C.P.; methodology, R.D., B.Y. and Z.W.; simulation, B.Y. and Z.W.; experimental investigation, R.D.; writing—original draft preparation, R.D.; writing—review and editing, Z.W. and C.P.; supervision, Z.W. and C.P. All authors contributed to the general discussions and have read and agreed to the published version of the manuscript.

Funding: This research was funded by European Union's Horizon 2020 research and innovation program, grant number 737164 (SUMCASTEC).

Data Availability Statement: The data are available from the corresponding author upon reasonable request.

Conflicts of Interest: The authors declare no conflict of interest.

References

1. Abbe, E. Beiträge zur Theorie des Mikroskops und der mikroskopischen Wahrnehmung. *Archiv. Mikrosk. Anat.* **1873**, *9*, 413–418. [CrossRef]
2. Novotny, L.; Hecht, B. *Principles of Nano-Optics*; Cambridge University Press: Cambridge, UK, 2012.
3. Pohl, D.W.; Denk, W.; Lanz, W. Optical stethoscopy: Image recording with resolution $\lambda/20$. *Appl. Phys. Lett.* **1984**, *44*, 651. [CrossRef]
4. Pendry, J.B. Negative Refraction Makes a Perfect Lens. *Phys. Rev. Lett.* **2000**, *85*, 3966. [CrossRef] [PubMed]
5. Shelby, R.A.; Smith, D.R.; Schultz, S. Experimental verification of a negative index of refraction. *Science* **2001**, *292*, 77–79. [CrossRef]
6. Smith, D.R.; Padilla, W.J.; Vier, D.C.; Nemat-Nasser, S.C.; Schultz, S. Composite medium with simultaneously negative permeability and permittivity. *Phys. Rev. Lett.* **2000**, *84*, 4184–4187. [CrossRef] [PubMed]
7. Fang, N.; Lee, H.; Sun, C.; Zhang, X. Sub-diffraction-limited optical imaging with a silver superlens. *Science* **2005**, *308*, 534–537. [CrossRef] [PubMed]
8. Jacob, Z.; Alekseyev, L.V.; Narimanov, E. Optical hyperlens: Far-field imaging beyond the diffraction limit. *Opt. Express* **2006**, *14*, 8247–8256. [CrossRef]
9. Liu, Z.; Lee, H.; Xiong, Y.; Sun, C.; Zhang, X. Far-field optical hyperlens magnifying sub-diffraction-limited objects. *Science* **2007**, *315*, 1686. [CrossRef]
10. Xiong, Y.; Liu, Z.; Sun, C.; Zhang, X. Two-dimensional Imaging by far-field superlens at visible wavelengths. *Nano Lett.* **2007**, *7*, 3360–3365. [CrossRef]
11. Dhama, R.; Rashed, A.R.; Caligiuri, V.; El Kabbash, M.; Strangi, G.; De Luca, A. Broadband optical transparency in plasmonic nanocomposite polymer films via exciton-plasmon energy transfer. *Opt. Express* **2016**, *24*, 14632–14641. [CrossRef] [PubMed]
12. Betzig, E.; Patterson, G.H.; Sougrat, R.; Lindwasser, O.W.; Olenych, S.; Bonifacino, J.S.; Davidson, M.W.; Lippincott-Schwartz, J.; Hess, H.F. Imaging intracellular fluorescent proteins at nanometer resolution. *Science* **2006**, *313*, 1642–1645. [CrossRef]
13. Wang, Z.; Guo, W.; Li, L.; Lukyanchuk, B.; Khan, A.; Liu, Z.; Chen, Z.; Hong, M. Optical virtual imaging at 50 nm lateral resolution with a white-light nanoscope. *Nat. Commun.* **2011**, *2*, 218. [CrossRef]
14. Darafsheh, A.; Limberopoulos, N.I.; Derov, J.S.; Walker, D.E., Jr.; Astratov, V.N. Advantages of microsphere-assisted super-resolution imaging technique over solid immersion lens and confocal microscopies. *Appl. Phys. Lett.* **2014**, *104*, 061117. [CrossRef]
15. Wang, F.F.; Liu, L.Q.; Yu, H.B.; Wen, Y.D.; Yu, P.; Liu, Z.; Wang, Y.C.; Li, W.J. Scanning superlens microscopy for non-invasive large field-of-view visible light nanoscale imaging. *Nat. Commun.* **2016**, *7*. [CrossRef]
16. Darafsheh, A.; Walsh, G.F.; Dal Negro, L.; Astratov, V.N. Optical super-resolution by high-index liquid-immersed microspheres. *Appl. Phys. Lett.* **2012**, *101*. [CrossRef]
17. Lee, S.; Li, L.; Wang, Z.; Guo, W.; Yan, Y.; Wang, T. Immersed transparent microsphere magnifying sub-diffraction-limited objects. *Appl. Opt.* **2013**, *52*, 7265–7270. [CrossRef]
18. Fan, W.; Yan, B.; Wang, Z.; Wu, L. Three-dimensional all-dielectric metamaterial solid immersion lens for subwavelength imaging at visible frequencies. *Sci. Adv.* **2016**, *2*, e1600901. [CrossRef]
19. Yue, L.Y.; Yan, B.; Wang, Z.B. Photonic nanojet of cylindrical metalens assembled by hexagonally arranged nanofibers for breaking the diffraction limit. *Opt. Lett.* **2016**, *41*, 1336–1339. [CrossRef] [PubMed]
20. Monks, J.N.; Yan, B.; Hawkins, N.; Vollrath, F.; Wang, Z.B. Spider Silk: Mother Nature's Bio-Superlens. *Nano Lett.* **2016**, *16*, 5842–5845. [CrossRef]
21. Wang, Z.; Lukyanchuk, B. Super-resolution imaging and microscopy by dielectric particle-lenses. In *Label-Free Super-Resolution Microscopy*; Astratov, V., Ed.; Springer: Berlin/Heidelberg, Germany, 2019.
22. Wang, Z.B.; Lukyanchuk, B.; Yue, L.Y.; Yan, B.; Monks, J.; Dhama, R.; Minin, O.V.; Minin, I.V.; Huang, S.M.; Fedyanin, A.A. High order Fano resonances and giant magnetic fields in dielectric microspheres. *Sci. Rep.* **2019**, *9*. [CrossRef]
23. Yue, L.Y.; Yan, B.; Monks, J.N.; Dhama, R.; Jiang, C.L.; Minin, O.V.; Minin, I.V.; Wang, Z.B. Full three-dimensional Poynting vector flow analysis of great field-intensity enhancement in specifically sized spherical-particles. *Sci. Rep.* **2019**, *9*. [CrossRef]
24. Wang, Z.B. Microsphere super-resolution imaging. In *Nanoscience*; Paual O Brien, J.T., Ed.; Royal Society of Chemistry: London, UK, 2016; Volume 3, pp. 193–210.
25. Luk'yanchuk, B.S.; Paniagua-Dominguez, R.; Minin, I.; Minin, O.; Wang, Z.B. Refractive index less than two: Photonic nanojets yesterday, today and tomorrow [Invited]. *Opt. Mater. Express* **2017**, *7*, 1820–1847. [CrossRef]
26. Kang, D.; Pang, C.; Kim, S.M.; Cho, H.S.; Um, H.S.; Choi, Y.W.; Suh, K.Y. Shape-Controllable Microlens Arrays via Direct Transfer of Photocurable Polymer Droplets. *Adv. Mater.* **2012**, *24*, 1709–1715. [CrossRef]
27. Darafsheh, A.; Guardiola, C.; Palovcak, A.; Finlay, J.C.; Cárabe, A. Optical super-resolution imaging by high-index microspheres embedded in elastomers. *Opt. Lett.* **2015**, *40*, 5–8. [CrossRef] [PubMed]
28. Yan, B.; Wang, Z.B.; Parker, A.L.; Lai, Y.K.; Thomas, P.J.; Yue, L.Y.; Monks, J.N. Superlensing microscope objective lens. *Appl. Opt.* **2017**, *56*, 3142–3147. [CrossRef] [PubMed]

Communication

Near- to Far-Field Coupling of Evanescent Waves by Glass Microspheres

Rayenne Boudoukha [1,2], Stephane Perrin [1,*], Assia Demagh [2], Paul Montgomery [1], Nacer-Eddine Demagh [2] and Sylvain Lecler [1]

[1] ICube Research Institute, CNRS—INSA Strasbourg—University of Strasbourg, 67412 Illkirch, France; rayenne.boudoukha@etu.unistra.fr (R.B.); paul.montgomery@unistra.fr (P.M.); sylvain.lecler@insa-strasbourg.fr (S.L.)

[2] Applied Optics Laboratory, Institute of Optics and Precision Mechanics, Ferhat Abbas University Setif1, 19000 Setif, Algeria; assiademagh@univ-setif.dz (A.D.); ndemagh@univ-setif.dz (N.-E.D.)

* Correspondence: stephane.perrin@unistra.fr

Abstract: Through rigorous electromagnetic simulations, the natural coupling of high-spatial-frequency evanescent waves from the near field to the far field by dielectric microspheres is studied in air. The generation of whispering gallery modes inside the microspheres is shown independently of any resonance. In addition, the conversion mechanism of these evanescent waves into propagating waves is analysed. This latter point leads to key information that allows a better physical understanding of the super-resolution phenomenon in microsphere-assisted microscopy where sub-diffraction-limit revolving power is achieved.

Keywords: microsphere; evanescent waves; propagation; whispering gallery modes; super-resolution

Citation: Boudoukha, R.; Perrin, S.; Demagh, A.; Montgomery, P.; Demagh, N.-E.; Lecler, S. Near- to Far-Field Coupling of Evanescent Waves by Glass Microspheres. *Photonics* **2021**, *8*, 73. https://doi.org/10.3390/photonics8030073

Received: 5 February 2021
Accepted: 4 March 2021
Published: 6 March 2021

Publisher's Note: MDPI stays neutral with regard to jurisdictional claims in published maps and institutional affiliations.

1. Introduction

In optical imaging, the spatial distribution of an object leads to a spatial modulation of the reflected (or transmitted) light. According to the diffraction theory of light, only a part of the electromagnetic waves having spatial frequencies smaller than $2/\lambda$ (λ, the wavelength of light) in air appears to be propagative and, thus, can be collected by an optical system. The high-spatial-frequency modulations remain localised around the object surface and are evanescent. With the aim to observe more and more details, various optical techniques have been developed in order to overcome the diffraction limit, such as near-field scanning optical microscopy [1], stimulated-emission-depletion microscopy [2] and photo-activated localisation microscopy [3], and the metamaterial-based superlens [4].

Respectively, in 2009 and in 2011, the ability of nanoscale hemispheres [5] and of microspheres [6] to perform super resolution imaging was also experimentally revealed. Fort this purpose, a dielectric spherical lens is deposited on the object surface to be tested. A magnified image carrying the fine details is then generated [7] in order to be collected by a microscope objective. Both make it possible to observe sub-diffraction-limit features while being label-free, scanning-free and easy-to-implement. Nevertheless, the magnification factor and the resolving power seem to be higher in microsphere-assisted microscopy than in nanolens-based microscopy. Indeed, the microspheres are able to improve the resolving power of an optical microscope by a factor of up to ×4 in air [6] and ×6 in immersion [8] while spherical nanolenses offer a magnification factor of 1.6 in air [5]. Convenient in biological imaging [9–11] and in optical profilometry [12], microsphere-assisted microscopy has thus become a powerful technique using a single microsphere or a matrix of microspheres [13].

Nowadays, the phenomenon behind the super resolution is not fully described [14]. It has yet been shown that the photonic jet cannot be interpreted as the point spread function of the imaging system [15,16], despite its promising performance in terms of forming a narrow focusing beam [17] for scanning optical imaging [18,19]. The role of coherence in the imaging process has been pointed out where two point objects need to be out-of-phase in order to be resolved [20]. The role of evanescent waves was first experimentally suggested through a decrease in resolution with an increase in the distance microsphere-sample [21] and more-recently confirmed using semi-immersed $BaTiO_3$ microspheres [22]. To our knowledge, the most advanced explanation has been proposed by Zhou et al. [23] where whispering-gallery modes excited inside the microsphere by high-spatial frequencies from the object make it possible to contribute to a virtual image resolution. However, the numerical demonstration was performed merely by implementing two point sources without indicating the specific contribution of the evanescent waves from the point sources to the image formation in the far field. In 2016, the possible refraction of evanescent waves by microspheres was mathematically demonstrated without any proposition of predictable imaging processing [24]. Therefore, a detailed description of the role of evanescent waves in the microsphere super-resolution imaging process is still expected. In this work, not only the natural coupling of evanescent waves by the glass microspheres, but also their conversion into propagating waves are studied, contributing to the clarification of the physical mechanism of super resolution in microsphere-assisted microscopy.

2. Method

To investigate the evanescent wave collection, a rigorous electromagnetic simulation of microsphere assisted microscopy was implemented in the visible range using a finite element method (COMSOL Multiphysics®). In the x-z plane, the 2D model consists of a glass microsphere deposited on a dielectric-air interface at z = 0 µm. The transparent microsphere has a refractive index of 1.5 and a diameter D. The surrounding medium of the microsphere is air and the refractive index of the dielectric substrate n_s is of 1.5. An oblique excitation plane wave with a wavelength of λ_0/n_s (λ_0 = 600 nm) allows the evanescent wave to be generated on the substrate interface using total internal reflection. The incident angle θ is of 1 rad which is larger than the critical angle ($\sin \theta_c = 1/n_s$). The incident coherent excitation wave has a unit electric field, giving a radiated power density $P_0 = \epsilon c |E|^2/2$ with ϵ the permittivity of the dielectric substrate. The evanescent wave thus propagates along the x-axis and exponentially decays along the z-axis. According to the complex Snell-Descartes's law, the wave vector components of the evanescent wave in air can be written as:

$$\vec{K} = \begin{cases} k_0\, n_s\, \sin\theta & \text{x axis} \\ i\, k_0\, \sqrt{n_s^2 \sin^2\theta - 1} & \text{z axis} \end{cases} \tag{1}$$

where k_0 is the wavenumber of the excitation plane wave in free-space. The imaginary unit i satisfies the relation $i^2 = -1$. Finally, perfectly matched layers surround the calculation area [25].

3. Results

Figure 1 shows the natural collection of the evanescent wave travelling along the interface by a 3-µm-diameter glass microsphere. The checked pattern in the dielectric substrate, in Figure 1a, results from interferences between the excitation plane wave and the wave reflected by the interface. Whispering gallery modes (WGMs) can be observed inside the microsphere. The highest peak amplitude of the WGM, equalling 3.7 V/m, occurs within the microsphere at a radius of around 1.4 µm.

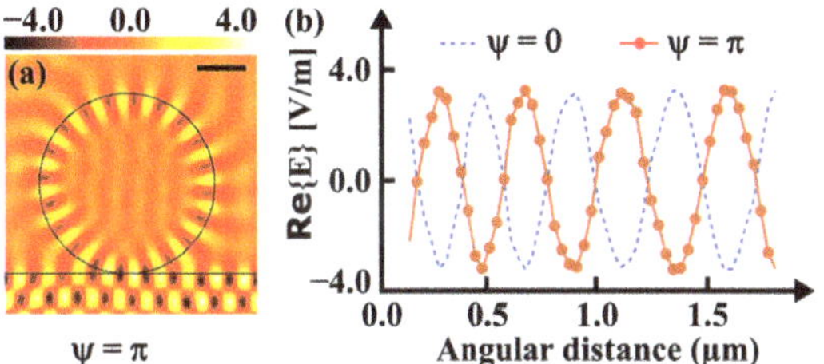

Figure 1. Collection of an evanescent wave by a glass microsphere (see Visualization 1 in Supplementary Materials). (**a**) Real part of the electric field distributions when the initial phase ψ of the excitation wave is of π. Black-scale bars represent 1 µm. (**b**) Radial section profiles at a radius of 1.4 µm of (**a**) as a function of the angular position when ψ equals 0 rad and π rad. λ_0 = 600 nm, θ = 1 rad, n_s = 1.5, D = 3 µm.

In order to emphasise the WGM generation from the evanescent wave and to confirm the coupling of the near-field wave by the microsphere, the initial phasor ψ of the excitation plane wave is swept from 0 to 2π. Increasing the offset ψ leads to a rotation of the WGMs (see Visualization 1 in Supplementary Materials). The rotational angle of the WGM patterns equals the initial phasor ψ. Indeed, as shown in Figure 1b, when ψ is of 0 rad and π rad, the angular sections of the electric field inside the sphere at a radius of 1.4 µm are also phase-shifted by π. Along the microsphere rim, the period of the WGM electric field equals 475 nm, i.e., the sub-wavelength period of the evanescent wave $\lambda_0/(n_s \times \sin\theta)$. At the radial position of 1.4 µm, the period of the WGM is smaller (around 445 nm). The peak amplitudes are, for their parts, not affected by the phase delay ψ (e.g., the highest peak amplitude of the WGM remains at around 3.7 V/m).

Figure 2 shows the ability of the 3-µm-diameter microsphere to collect the evanescent waves according to its axial position above the interface.

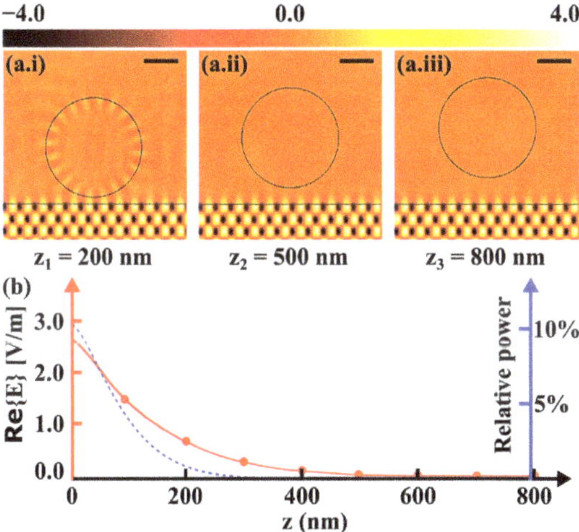

Figure 2. Collection efficiency of an evanescent wave by a glass microsphere at an axial gap with the interface. (**a.i**), (**a.ii**) and (**a.iii**) Real parts of the electric field distributions when the microsphere is positioned at a distance of 200 nm, 500 nm and 800 nm from the interface, respectively. Black-scale bars represent 1 µm. (**b**) Real-part peak amplitude of the WGM electric field along the microsphere rim and relative power transmitted by the microsphere in the far field as a function of the z distance. λ_0 = 600 nm, θ = 1 rad, ψ = 0 rad, n_s = 1.5, D = 3 µm.

Increasing the air gap allows not only to attest the evanescent wave coupling by the microsphere, but also to predict the longest acceptable distance, thus making it possible to perform contact-less measurements in microsphere-assisted microscopy. Indeed, the electric-field amplitude maximum of the WGM exponentially decreases according to the axial distance. In this case, the coupling efficiency of the evanescent wave by the microsphere appears to be low beyond a distance z = 400 nm. In Figure 2b, the far-field relative radiated power was estimated by considering the incident power P_0 through the radius of the estimated interaction area [21], i.e., $\sqrt{2\lambda_0 h}$ with h, the penetration depth measuring around 125 nm in this configuration $(1/|k_z|$, see Equation (1)). The optical power emanating from the top of the microsphere is not null and decreases as distance increases (e.g., 0.9% P_0 at z = 200 nm and 0.1% P_0 at z = 400 nm). In this case, the distance of 400 nm can be seen as the depth of field of the imaging technique.

A slight variation of the microsphere diameter obviously leads to modifications of the excited WGMs. Figure 3a highlights the peak amplitude fluctuations of the WGMs along the microsphere rim by varying the diameter from 3.0 µm to 3.3 µm. The microsphere is in contact with the interface. The amplitude of the electric field progresses following a periodic wave as a function of the diameter. In this case, periodic resonances occur when the microsphere diameter measures 3.03 µm and 3.17 µm which correspond to multiples of 475.3 nm/π. As a matter of fact, the multiple is the number of radial periods of the WGM, i.e., 20 and 21 periods.

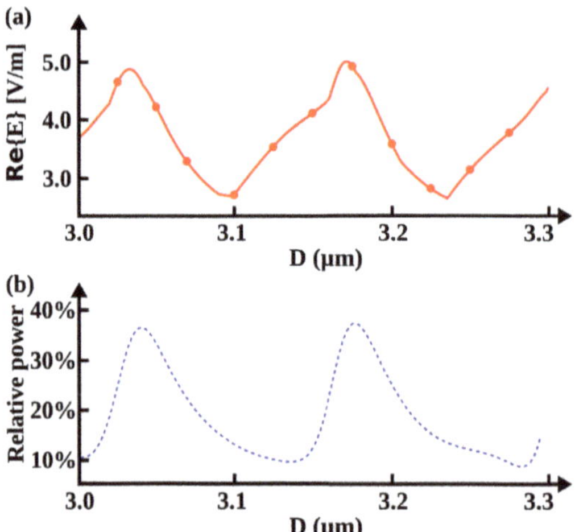

Figure 3. Influence of the microsphere diameter (**a**) on the peak amplitude of the real part of the electric field along the microsphere rim and (**b**) on the transmitted power. λ_0 = 600 nm, θ = 1 rad, ψ = 0 rad, n_s = 1.5.

Close resonances may appear but are not required to benefit from the evanescent wave coupling effect. As shown in Figure 3a, the resonances only increase the coupling efficiency. In addition, simulations were made (in the contact mode) in order to quantify the difference of the power transmitted in the far field by the microsphere for a resonant and a non-resonant WGM (see Figure 3b). n the far field, the transmitted power is increased by a factor of nearly 4 (at D = 3.14 µm, 9.8% P_0, and at D = 3.18 µm, 37.5% P_0). It can be observed that far field power peaks occur when the WGM electric field is maximum (at D = 3.03 µm and D = 3.17 µm). For the minima, this correlation is not found, indicating that the WGM generation is not the only phenomenon behind far field propagation.

Without the microsphere, the evanescent wave would propagate along the dielectric-air interface on the x-axis, i.e., a microscope objective alone would not be able to collect the high-spatial-frequency optical signal. However, by introducing the glass microsphere, the evanescent wave is not only coupled, but also appears to be propagated in the far field. Propagating-spherical waves emanate from the microsphere, demonstrating its ability to convert a near-field wave into a far-field wave. In experiments, these propagating waves are collected by a microscope objective. Figure 4 exposes this phenomenon in air when the period of the evanescent wave is 475 nm, 355 nm and 240 nm. For this purpose, the refractive index n_s of the dielectric substrate is varied (such refractive indices cannot be achieved in reality). Increasing the refractive index n_s also makes the periods of WGMs narrower. The optical power transmitted by the microsphere in the far field is of 10.9% P_0, 0.6% P_0 and 7.2 $\times 10^{-5}$% P_0 when n_s is of 1.5, 2.0 and 3.0, respectively. In Figure 4c,d, the amplitude and the intensity ranges are saturated in order to emphasise the relatively-low amplitude values of the propagating wave.

Thus, the ability of the microsphere to collect the highest spatial frequencies appears to be more complex. Recently, Lai et al. have experimentally highlighted that using a larger microsphere would lead to an increased efficiency of the transmitted intensity, but would reduce the resolving power [26]. In addition, a broadband light source usually illuminates the microsphere in experiments (instead of a coherent illumination). Due to the equivalence between the sphere diameter increase and the wavelength decrease in electromagnetism, Figure 3 shows that several nearby resonances will occur together using a broadband light source. Thus, each spectral component (or wavelength-dependent evanescent wave) would contribute to the conversion efficiency by the microsphere with more-or-less ability, yielding an increase in propagating-wave intensities. Moreover, to improve the signal-to-noise ratio of the imaging contrast, plasmonic-based illumination has been suggested [27]. Finally, we can notice that increasing the spatial frequency of the evanescent wave leads to a decrease in the attenuation distance of the near-field wave and, thus, to a decrease in the axial distance along which evanescent waves can be collected (e.g., 400 nm when $n_s = 1.5$).

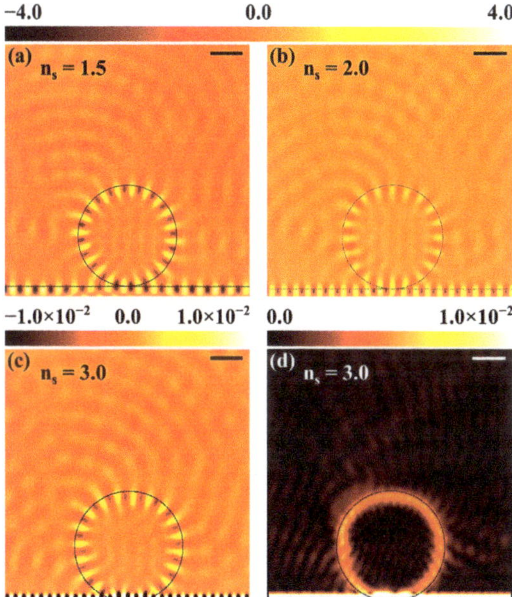

Figure 4. Conversion of an evanescent wave in a propagative wave by a glass microsphere. Real parts of the electric field distributions when (**a**) $n_s = 1.5$, (**b**) $n_s = 2.0$ and (**c**) $n_s = 3.0$. (**d**) Absolute value of the complex electric field distribution when $n_s = 3.0$. Scale bars represent 1 μm. $\lambda_0 = 600$ nm, $\theta = 1$ rad, $\psi = 0$ rad, D = 3 μm.

Figure 4 also shows that the outgoing propagating waves do not come from everywhere on the microsphere, but from specific points on its surface. This is due to the fact that the sphere interaction with the surface breaks the axial symmetry of the WGM [28]. This observation is important. More than a simple power collection (as in NSOM), this demonstrates the necessary relation between the spatial frequency and initial phasor ψ coming from the object plane and the spatial emitting position, a prerequisite for achieving full-field imaging.

In experiments, immersion of high-refractive-index microspheres may be required such as in biological imaging [8,29] or may offer the possibility of manipulating microspheres above the sample [30]. Working in immersion increases not only the resolving power, but also the imaging contrast. Indeed, a more efficient ability of microspheres in immersion to collect high-frequency waves occurs in immersion, enhancing the experimental conditions and performance. The gap distance would appear to be reduced in immersion.

4. Conclusions

The phenomenon behind super resolution in microsphere-assisted microscopy has been investigated. Based on a finite element method, rigorous electromagnetic simulations in the visible range have shown not only the natural collection of evanescent waves by dielectric microspheres through WGMs, but also their conversion into propagating waves. The near-field high-spatial frequencies of an object are naturally coupled without requiring resonances contributing to the explanation of the sub-diffraction-limit imaging ability of glass microspheres. Furthermore, it aims at proposing a comprehensive physical imaging mechanism. An analogy with the principle of near-field scanning optical microscopy can thus be made with enhancements such as the wide-field measurement. In future studies, a complete description of the imaging process using will be implemented in order to quantify the resolving power according to geometrical and optical parameters. Moreover, simulations through microspheres having a diameter of larger than 3 µm (as is often the case in experiments) and the influence of WGMs on the spectrum of the broadband light source will be considered.

Supplementary Materials: The following are available online at www.mdpi.com/xxx/s1.

Author Contributions: Methodology, R.B. and S.L.; software, R.B. and S.L.; validation, S.P. and S.L.; formal analysis, R.B.; investigation, R.B., S.P. and S.L.; resources, S.L.; data curation, R.B. and S.P.; writing—original draft preparation, R.B.; writing—review and editing, S.P., A.D., P.M., N.-E.D. and S.L.; visualization, R.B. and S.P.; supervision, A.D., P.M., N.-E.D. and S.L.; project administration, N.-E.D. and S.L.; funding acquisition, N.-E.D. and S.L. All authors have read and agreed to the published version of the manuscript.

Funding: SATT Conectus Alsace; PHC TASSILI Cooperation Program; University of Strasbourg.

Institutional Review Board Statement: Not applicable.

Informed Consent Statement: Not applicable.

Acknowledgments: The authors thank Amandine Elchinger for her useful comments in the editing of the manuscript.

Conflicts of Interest: The authors declare no conflict of interest.

References

1. Pohl, D.W.; Courjon, D. *Near Field Optics*; Springer: Amsterdam, The Netherlands, 1993.
2. Hell, S.W.; Wichmann, J. Breaking the diffraction resolution limit by stimulated emission: Stimulated-emission-depletion fluorescence microscopy. *Opt. Lett.* **1994**, *19*, 780–782. [CrossRef]
3. Betzig, E.; Patterson, G.H.; Sougrat, R.; Lindwasser, O.W.; Olenych, S.; Bonifacino, J.S.; Davidson, M.W.; Lippincott-Schwartz, J.; Hess, H.F. Imaging intracellular fluorescent proteins at nanometer resolution. *Science* **2006**, *313*, 1642–1645. [CrossRef]
4. Smith, D.R.; Schurig, D.; Rosenbluth, M.; Schultz, S.; Ramakrishna, S.A.; Pendry, J.B. Limitations on subdiffraction imaging with a negative refractive index slab. *Appl. Phys. Lett.* **2003**, *82*, 1506–1508. [CrossRef]
5. Lee, J.Y.; Hong, B.H.; Kim, W.Y.; Min, S.K.; Kim, Y.; Jouravlev, M.V.; Bose, R.; Kim, K.S.; Hwang, I.-C.; Kaufman, L.J.; Wong, C.W.; Kim, P.; Kim, K.S. Near-field focusing and magnification through self-assembled nanoscale spherical lenses. *Nature* **2009**, *460*, 498–501. [CrossRef]

6. Wang, Z.; Guo, W.; Li, L.; Luk'yanchuk, B.; Khan, A.; Liu, Z.; Chen, Z.; Hong, M. Optical virtual imaging at 50 nm lateral resolution with a white-light nanoscope. *Nat. Commun.* **2011**, *2*, 218. [CrossRef] [PubMed]
7. Perrin, S.; Li, H.; Lecler, S.; Montgomery, P. Unconventional magnification behaviour in microsphere-assisted microscopy. *Opt. Laser Technol.* **2019**, *114*, 40–43. [CrossRef]
8. Darafsheh, A. Influence of the background medium on imaging performance of microsphere-assisted super-resolution microscopy. *Opt. Lett.* **2017**, *42*, 735–738. [CrossRef] [PubMed]
9. Li, L.; Guo, W.; Yan, Y.; Lee, S.; Wang, T. Label-free super-resolution imaging of adenoviruses by submerged microsphere optical nanoscopy. *Light Sci. Appl.* **2013**, *2*, e104. [CrossRef]
10. Wang, F.; Liu, L.; Yu, H.; Wen, Y.; Yu, P.; Liu, Z.; Wang, Y.; Li, W.J. Scanning superlens microscopy for non-invasive large field-of-view visible light nanoscale imaging. *Nat. Commun.* **2016**, *7*, 13748. [CrossRef] [PubMed]
11. Perrin, S.; Li, H.; Badu, K.; Comparon, T.; Quaranta, G.; Messaddeq, N.; Lemercier, N.; Montgomery, P.; Vonesch, J.-L.; Lecler, S. Transmission microsphere-assisted dark-field microscopy. *Phys. Status Solidi RRL* **2019**, *13*, 1800445. [CrossRef]
12. Perrin, S.; Donie, Y.J.; Montgomery, P.; Gomard, G.; Lecler, S. Compensated Microsphere-Assisted Interference Microscopy. *Phys. Rev. Appl.* **2020**, *13*, 014068. [CrossRef]
13. Allen, K.W.; Farahi, N.; Li, Y.; Limberopoulos, N.I.; Walker, D.E.; Urbas, A.M.; Astratov, V.N. Overcoming the diffraction limit of imaging nanoplasmonic arrays by microspheres and microfibers. *Opt. Express* **2015**, *23*, 24484–24496. [CrossRef] [PubMed]
14. Duan, Y.; Barbastathis, G.; Zhang, B. Classical imaging theory of a microlens with super-resolution. *Opt. Lett.* **2013**, *38*, 2988–2990. [CrossRef]
15. Maslov, A.; Astratov, V. Resolution and reciprocity in microspherical nanoscopy: Point-spread function versus photonic nanojets. *Phys. Rev. Appl.* **2019**, *11*, 064004. [CrossRef]
16. Lecler, S.; Perrin, S.; Leong-Hoi, A.; Montgomery, P. Photonic jet lens. *Sci. Rep.* **2019**, *9*, 4725. [CrossRef] [PubMed]
17. Darafsheh, A. Photonic nanojets and their applications. *J. Phys. Photonics* **2021**, *3*, 022001. [CrossRef]
18. Yi, K.J.; Wang, H.; Lu, Y.F.; Yang, Z.Y. Enhanced Raman scattering by self-assembled silica spherical microparticles. *J. Appl. Phys.* **2007**, *101*, 063528. [CrossRef]
19. Tehrani, K.F.; Darafsheh, A.; Phang, S.; Mortensen, L.J. Resolution enhancement of 2-photon microscopy using high-refractive index microspheres. *Proc. SPIE* **2018**, *10498*, 1049833.
20. Kassamakov, I.; Lecler, S.; Nolvi, A.; Leong-Hoi, A.; Montgomery, P.; Haeggstrom, E. 3D super-resolution optical profiling using microsphere enhanced Mirau interferometry. *Sci. Rep.* **2017**, *7*, 3683. [CrossRef] [PubMed]
21. Lin, Q.; Wang, D.; Wang, Y.; Rong, L.; Zhao, J.; Guo, S.; Wang, M. Super-resolution imaging by microsphere-assisted optical microscopy. *Opt. Quantum Electron.* **2016**, *48*, 557. [CrossRef]
22. Yang, S.; Ye, Y.-H.; Shi, Q.; Zhang, J. Converting Evanescent Waves into Propagating Waves: The Super-Resolution Mechanism in Microsphere-Assisted Microscopy. *J. Phys. Chem. A C* **2020**, *124*, 25951–25956. [CrossRef]
23. Zhou, S.; Deng, Y.; Zhou, W.; Yu, M.; Urbach, H.P.; Wu, Y. Effects of whispering gallery mode in microsphere super-resolution imaging. **2017**, *123*, 236 [CrossRef]
24. Ben-Aryeh, Y. Increase of resolution by use of microspheres related to complex Snell's law. *JOSA A* **2016**, *33*, 2284–2288. [CrossRef] [PubMed]
25. Berenger, J.-P. A perfectly matched layer for the absorption of electromagnetic waves. *J. Comput. Phys.* **1994**, *114*, 185–200. [CrossRef]
26. Lai, H.S.S.; Wang, F.; Li, Y.; Jia, B.; Liu, L.; Li, W.J. Super-resolution real imaging in microsphere-assisted microscopy. *PLoS ONE* **2016**, *11*, e0165194. [CrossRef]
27. Astratov, V.N.; Brettin, A.; Abolmaali, F.; Poffo, L.; Maslov, A.V. Plasmonics and Super resolution in Microspherical Nanoscopy. In Proceedings of the 20th International Conference on Transparent Optical Networks (ICTON), Bucharest, Romania, 1–5 July 2018; pp. 1–4.
28. Mollaei, M.; Simovski, C. Dual-metasurface superlens: A comprehensive study. *Phys. Rev. B* **2019**, *100*, 205426. [CrossRef]
29. Darafsheh, A.; Walsh, G.F.; Negro, L.D.; Astratov, V.N. Optical super-resolution by high-index liquid-immersed microspheres. *Appl. Phys. Lett.* **2012**, *101*, 141128. [CrossRef]
30. Zhang, T.; Li, P.; Yu, H.; Wang, F.; Wang, X.; Yang, T.; Yang, W.; Li, W.J.; Wang, Y.; Liu, L. Fabrication of flexible microlens arrays for parallel super-resolution imaging. *Appl. Surf. Sci.* **2020**, *504*, 144375. [CrossRef]

photonics

Article

Near-Field Light-Bending Photonic Switch: Physics of Switching Based on Three-Dimensional Poynting Vector Analysis

Liyang Yue [1,*], Zengbo Wang [1], Bing Yan [1], Yao Xie [2], Yuri E. Geints [3], Oleg V. Minin [4] and Igor V. Minin [4]

1 School of Computer Science and Electronic Engineering, Bangor University, Dean Street, Bangor, Gwynedd LL57 1UT, UK; z.wang@bangor.ac.uk (Z.W.); b.yan@bangor.ac.uk (B.Y.)
2 School of Physics and Optoelectronic Engineering, Xidian University, No. 2 South Taibai Road, Xi'an 710071, China; yxie_100@stu.xidian.edu.cn
3 V. E. Zuev Institute of Atmospheric Optics, Siberian Branch, Russian Academy of Sciences, 1 Zuev Square, 634021 Tomsk, Russia; ygeints@iao.ru
4 Nondestructive Testing School, Tomsk Polytechnic University, 36 Lenin Avenue, 634050 Tomsk, Russia; ovminin@tpu.ru (O.V.M.); prof.minin@gmail.com (I.V.M.)
* Correspondence: l.yue@bangor.ac.uk

Abstract: Photonic hook is a high-intensity, bent light focus with a proportional curvature to the wavelength of the incident light. Based on this unique light-bending phenomenon, a novel near-field photonic switch by means of a right-trapezoid dielectric Janus particle-lens embedded in the core of a planar waveguide is proposed for switching the photonic signals at two common optical communication wavelengths, 1310 nm and 1550 nm, by using numerical simulations. The signals at these two wavelengths can be guided to different routes according to their oppositely bent photonic hooks to realise wavelength selective switching. The switching mechanism is analysed by an in-house developed three-dimensional (3D) Poynting vector visualisation technology. It demonstrates that the 3D distribution and number of Poynting vector vortexes produced by the particle highly affect the shapes and bending directions of the photonic hooks causing the near-field switching, and multiple independent high-magnitude areas matched by the regional Poynting vector streamlines can form these photonic hooks. The corresponding mechanism can only be represented by 3D Poynting vector distributions and is being reported for the first time.

Keywords: photonic hook; optical switch; Poynting vector; vortex

Citation: Yue, L.; Wang, Z.; Yan, B.; Xie, Y.; Geints, Y.E.; Minin, O.V.; Minin, I.V. Near-Field Light-Bending Photonic Switch: Physics of Switching Based on Three-Dimensional Poynting Vector Analysis. *Photonics* **2022**, *9*, 154. https://doi.org/10.3390/photonics9030154

Received: 29 January 2022
Accepted: 3 March 2022
Published: 4 March 2022

Publisher's Note: MDPI stays neutral with regard to jurisdictional claims in published maps and institutional affiliations.

1. Introduction

Switch is the most basic component to make, break, or divert the connection from one conductor to another in a circuit. The conventional electronic switch is a solid-state mechanical device, though the sensors can be used to automate it in certain working environments. The concept of the switch can be used for signal processing and other applications of modern communication technologies, and activation/deactivation of signals and data routing are the major functions of a switch, sometimes called logic signal switch, in here [1]. With the development of data technology and mobile computing, the optical signal in the form of light replaced the electronic signal to become the mainstream method of data transmission with the aid of optical fibres [2,3]. For this reason, a photonic switch is developed to control the light (optical signals) propagation with high efficiency, and its switching time is much faster than that of a conventional electronic switch, due to the unparalleled speed of light [4,5]. In this field, wavelength switching is a technology used in optical communication to select individual wavelengths of light and forward the selected input light to the separate paths for specific data routing [6]. Meanwhile, many optical fibres are optimised for the optical signals at wavelengths of 1310 nm and 1550 nm due to their lower losses in the glass fibre [7]. Wavelength division multiplexing (WDM) and

Dense WDM (DWDM) are the technologies that enable a number of separate optical signals with different wavelengths to be transmitted in an optical fibre [8,9], but the sizes of these devices limit their applications for the future photonic integrated circuits and lab-on-a-chip devices. As an alternative, the planar dielectric waveguide, in the form of a dielectric film (core) sandwiched between cladding layers, better applies to these integrated optoelectronic devices due to a relatively simple and compact structure and the consequent production advantages [10,11]. Theoretically, the optical signal can be fast and accurately transmitted in the core of a planar waveguide relying on the total internal reflection occurring at the interface between the core and cladding [12].

In this paper, a novel all-dielectric photonic switch is realised by the near-field light-bending effect of photonic hook which is a curved and localised, high-intensity light beam focused by a dielectric micro-particle [13]. When it was invented, it was the only other instance of artificially bent light apart from the Airy beam [13,14]. Meanwhile, it is highly related to the concept of photonic jet or can be directly understood as a photonic jet asymmetrically shifting in optical phase due to difference in phase velocity caused by regional refractive-index contrast. Typically, it is produced by the light focusing through a dielectric right-trapezoid Janus microparticle with the asymmetry in geometrical shape and internal refractive index from its prism end [14–16]. This wavefront curvature normally appears in the shadow direction of the particle and significantly influence the electric field distribution in that area. Minin et al. experimentally generated a photonic hook in terahertz (THz) band for the first time using a relatively simple experimental setup, meanwhile proving the proportional relationship between the wavelength of incident light and curvature of photonic hook [15]. Based on this finding, Geints et al. raised the proofs-of-concept of similar photonic switches, and herein, photonic hooks could function like switching channels after passing several differently configured dielectric particles, such as the flipped prism, Janus bar, and off-axis Fresnel zone plate [17]. Moreover, most photonic hooks are graphically represented by the distributions of electric field and Poynting vector on the median section. This two-dimensional (2D) plot is sufficient to characterise the complete shape and curvature of a photonic hook; however, it is difficult to analyse the light-bending process and the physics behind the phenomenon because the influence of the photonic energy that flows into the median section from other planes cannot be quantified, even visualised [18]. Yue et al. developed a 3D mapping technology to track the power flow (Poynting vectors) of a photonic jet focused by a spherical mesoscale micro-particle, and the circulation and convergence of 3D Poynting vectors inside the dielectric particle were demonstrated for the first time [19]. The same technology is used in this study to solve the above issue.

In this article, we propose a novel photonic hook light-bending optical switch by means of a conventional right-trapezoid-shaped dielectric particle embedded in the core of a planar waveguide. The operation of signal switching at two communication wavelengths, $\lambda = 1310$ and 1550 nm, are numerically simulated. Its setup and operating wavelengths are designed for the applications of optical communication and are different from those in [17]. The separation of switching channels for the signals at these two wavelengths is demonstrated by the overall distribution of the electric field and profiles of the generated photonic hooks. The corresponding simulation data is analysed by 3D Poynting vector visualisation technology to deeply investigate the physics and mechanism of switching.

2. Results and Discussion

2.1. Concept and Modelling

The numerical model of the proposed photonic switch is built in the commercial finite integral technique (FIT) software package, CST Microwave Studio (CST), which is widely used for design and analysis of nanophotonic devices [19,20]. A right-trapezoid-shaped particle with the refractive index of $n = 2$, which can be possibly constituted by the dielectric materials with the similar refractive index, e.g., Boron nitride (BN), MICROPOSIT® S1800 photoresist (S1800), etc. (n ranges 1.9–2.1 in near-infrared band depending on thickness and structure), is implanted in the facet of the quartz core of a planar waveguide with $n = 1.44$ [21–23]. The difference of the refractive indexes for the particle material at the wavelengths of 1310 nm and 1550 nm is neglected to simplify and standardise the models for comparison purposes. The assumed planar waveguide is able to ideally approximate the plane wave propagation and eliminate the interference for the transmitted light at wavelengths of 1310 nm and 1550 nm, and the implanted particle with a small relative dimension to the core ($d/D \leq 0.08$, where d and D are the particle size and thickness of the core of the planar waveguide, respectively) [24,25]. The sketch of this design is shown in Figure 1a I and the diagrams of Figure 1a II and III illustrate the effect of a trapezoid particle switching the incident lights with short and long wavelengths which are 1310 nm and 1550 nm in this case. As a discussion of the proof of concept, the proposed photonic hook switch could be fabricated using the technologies of plano-convex-microsphere (PCM) lens laser nano-marking and two-photon polymerization (TPP) 3D printing (resolution of fs laser wavelength) with two example materials—BN and S1800, respectively—aided by the integrated high-precision piezo positioning stages and femtosecond (fs) lasers (positioning resolution of the stage: 1 nm; minimum corner radius of the structure: ~100 nm) [26,27]. A PCM lens is a dielectric lens made up of a high-index microsphere integrated with a plano-convex lens, which can deliver a high patterning resolution that is smaller than the diffraction limit for a precision machining of the facet of the planar waveguide core in a trapezoid shape by a fs laser [26]. TPP 3D printing is an additive fabrication technology to create micro/nano features relying on the photoresist solidification only occurring at the fs laser focus [27]. It can in situ fill the micro/nanosized photonic wires of dielectric material layer by layer in the cavity created by PCM lens marking technology with a high degree of design freedom.

For the simulation setup in CST, the core material with the refractive index of $n = 1.44$ is used to encapsulate the trapezoid particle as long as the flat end facing the background medium of air ($n = 1$) with the open boundary condition (effect of Perfectly matched layers, PML) along with x, y, and z directions to simulate the planar waveguide transmission and photonic hook switching in the air. The incident plane wave is y-polarized and propagates along z axis with the amplitude of 1 v/m. The trapezoid particle is placed as the prism encountering the light in z direction. The short base, long base, height, thickness, and prism angles of the trapezoid are 2.79 μm, 3.72 μm, 2.79 μm, 2.79 μm, and 71.57° in this design, respectively. Furthermore, the frequency domain solver and tetrahedral meshing with the average edge length of 188 nm (min. and max. edge lengths are 33 nm and 339 nm, respectively; the approximate level of $\lambda/7$) are used in the model for adaptive refinement of the simulation accuracy with the autonomous adjustment of mesh sizes on the surfaces and edges of the geometries. The modelling diagram is shown in Figure 1b. The data of the Poynting vector in the modelling space generated by CST is exported and then independently processed by a MATLAB® programme to plot the contour of the Poynting vector magnitudes in a certain plane and a full 3D distribution of the Poynting vector streamlines entering the particle.

2.2. Switching and 2D Distributions

The dielectric particle in the proposed photonic switch design is individually irradiated by the plane-wave lights (optical signals) at wavelengths of 1310 nm and 1550 nm, and Figure 2 shows the corresponding distributions of the squared total magnitudes of electric field ($|E|^2$, E is electric field) and real part of complex Poynting vectors and their directions (unit: V·A/m^2) [28]. The distributions of $|E|^2$ field in Figure 2a,b demonstrate a clear separation of switching paths. In Figure 2a, the wavefront of 1310 nm light develops in a route approaching the side of the trapezoid long base, which forms a downward photonic hook. By contrast, the development of the photonic hook at the wavelength of 1550 nm is different from that for 1310-nm-wavelength light. It propagates in an almost straight path with a slightly upward curvature in the middle. The optical signals with the same wavelengths should be switched to the routes in accordance with the photonic hook curvatures, and the distance between two routes in the shadow area of the particle can provide the dimensional tolerance for the device fabrication. Theoretically, two ports can be set up in the middle of the photonic hook routes to separately receive the corresponding signals.

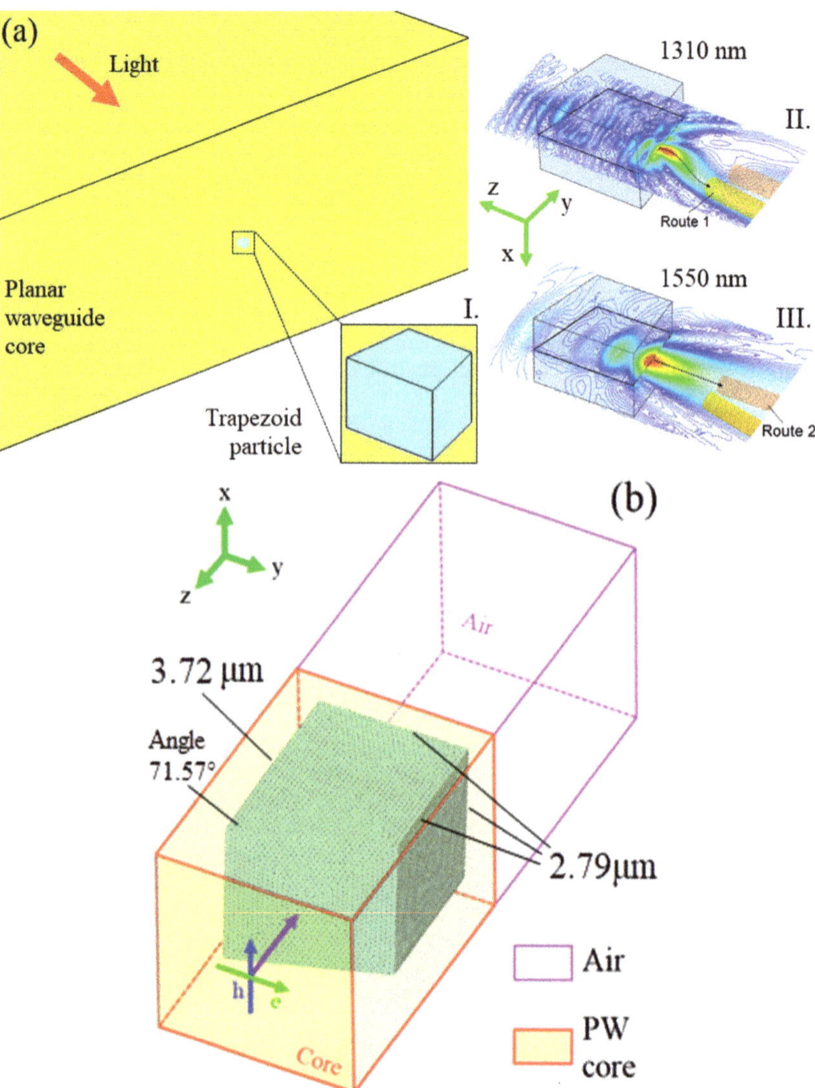

Figure 1. (**a**) I, diagram of photonic switch design; II and III, principle of the proposed photonic switching at two different wavelengths (**b**) CST modelling diagram.

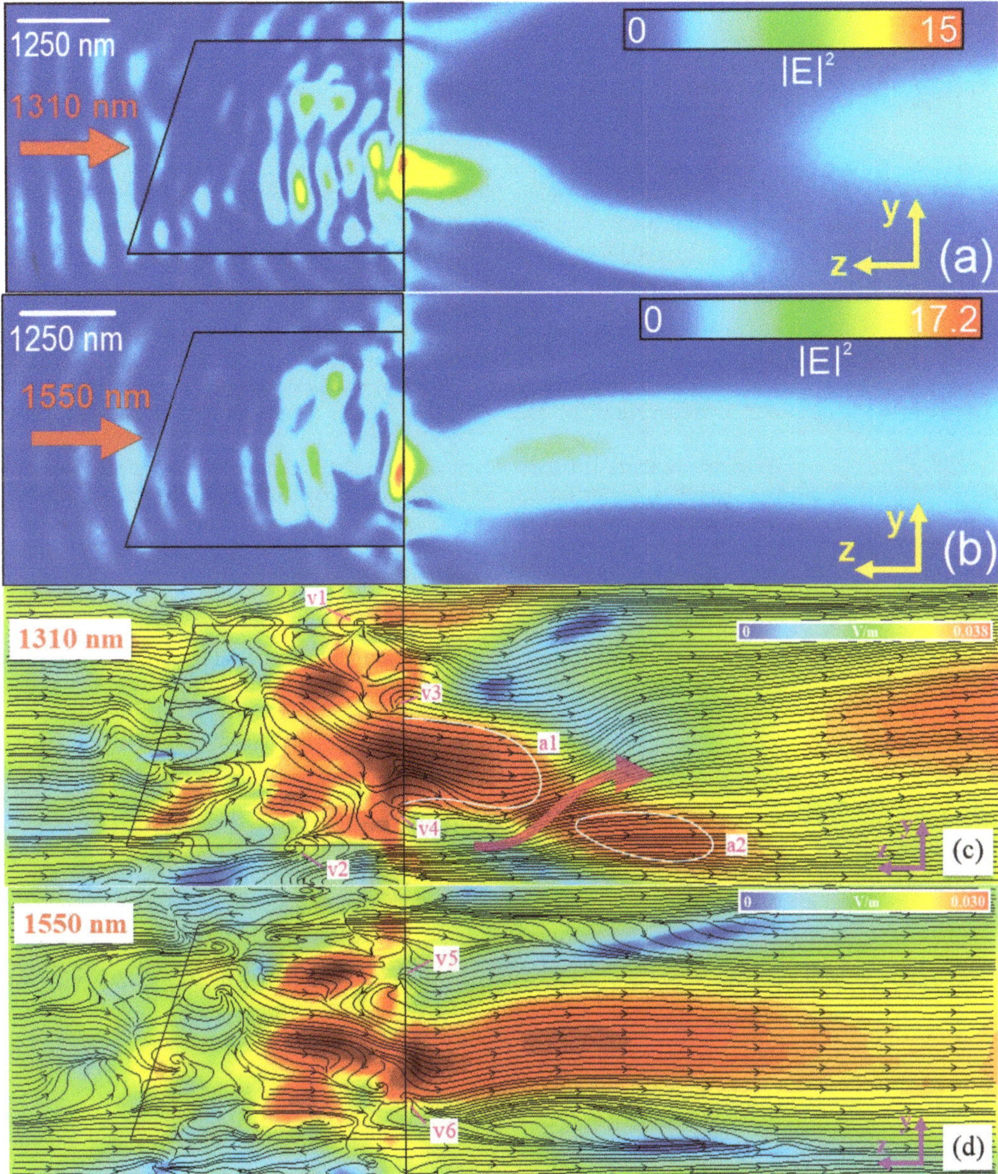

Figure 2. The distributions of the squared total magnitude of electric field ($|E|^2$) for the irradiations with wavelengths of 1310 nm (**a**) and 1550 nm (**b**). Two-dimensional Poynting vector distributions for the irradiations with wavelengths of 1310 nm (**c**) and 1550 nm (**d**). The dimensions of the trapezoid particle are short base, 2.79 µm; long base, 3.72 µm; height, 2.79 µm; thickness, 2.79 µm; and prism angle, 71.57°.

Figure 2c,d demonstrate the 2D Poynting vector distributions on the median section for the models irradiated by the optical signals with wavelengths of 1310 nm and 1550 nm, respectively. The asymmetric vortexes of Poynting vector streamlines are shown in the dielectric particles in both figures. It is known that the photonic energy is able to flow in and out through these vortexes representing the singularities of Poynting vectors possibly associated with topological charge [29–31]. Consequently, the layout of these vortexes can influence the flow direction and intensity of Poynting vectors, especially in the areas which are close to the particle boundaries. In Figure 2c for 1310 nm optical signal, the prism of the dielectric particle leads the difference of phase velocity in the particle and generates four vortexes at the output end of the particle, marked as v1, v2, v3, and v4 in the figure. Referring to the contour and flows of Poynting vectors in Figure 2c, it is shown that two saddle points, outward spiral v1 and inward spiral v2, are in high-intensity and low-intensity areas, respectively. This means that the photonic power flow passes into the median plane at v1 position in the upper half of the particle, meanwhile exiting the plane at v2 position in the lower half. The corresponding power flow exchange and influences of vortexes v3 and v4 around the output area of the particle make the flows of Poynting vectors in the generated photonic hook downward in general. However, it is noted that the photonic hook of 1310-nm-wavelength irradiation is divided into two high-magnitude areas (marked as a1 and a2) by a Poynting vector flow represented by a pink arrow in Figure 2c. This separation is unusual and rare in the discovered photonic hooks and different from the photonic hook in Figure 2a for the squared electric field distribution. Rather than the downward photonic hook for the optical signal of 1310 nm wavelength, a slightly upward photonic hook is created by the same trapezoid particle for the incidence of 1550-nm-wavelength light. Two outward vortexes, v5 and v6 in Figure 2d, shape the flow distribution and produce two narrow low-magnitude regions compressing the high-magnitude area of the photonic hook in the middle.

2.3. 3D Poynting Vector Analysis

The physical mechanism behind the switching of the optical signals at wavelengths of 1310 nm and 1550 nm and their field distributions in Figure 2 is further investigated by the 3D Poynting vector visualisation analysis [19]. Compared to 2D analysis, 3D analysis will provide additional information on light flows between different cross sections and also the root of formation of optical vortexes and other singular points. All Poynting vectors entering the dielectric particle from the prism end are tracked and then visualised by an in-house developed MATLAB® programme. Due to a large amount of Poynting vector data imported in the programme, the plotted Poynting vector streamlines are marked in different colours to indicate the regions where they are tracked in the model. The colour codes are distributed along the y axis to equally divide the whole modelling space into three sections: the blue lines for the long-base section ($-y$ direction), red lines for the middle section (around 0 position on the y axis), and green lines for the short-base section ($+y$ direction). An overview of the 3D Poynting vector distribution for the model of 1310-nm-wavelength irradiation is shown in Figure 3a. All Poynting vector streamlines flow upwards from the prism end to the flat facet of the particle (vector arrows are hidden due to a high density of streamlines); however, the presented features of 3D distribution are significantly different from those in the 2D format shown in Figure 2. The Poynting vector streamlines can continuously and freely develop in multiple directions and finally create a tangled-roots-like structure. Figure 3b as a close-up view of the same model demonstrates the locations of the vortexes with the insets of two important ones, v1 and v2, also indicated in Figure 2. Except for the different spin directions of v1 and v2 shown in Figure 2, 3D visualisation of Poynting vectors can graphically represent the orientations of v1 and v2 and fully indicate this difference in Figure 3b. It is shown that the centre and spin-orbit of v1 is approximately perpendicular to the z axis and faces upward in -z direction, meanwhile, the same features are generally in the direction of 45° to the x axis for v2. In addition, the inner and outer circles of v1 are in red and green, respectively, which means the photonic energy can flow to the middle and upper parts of the particle and modelling space through this vortex using multiple routes.

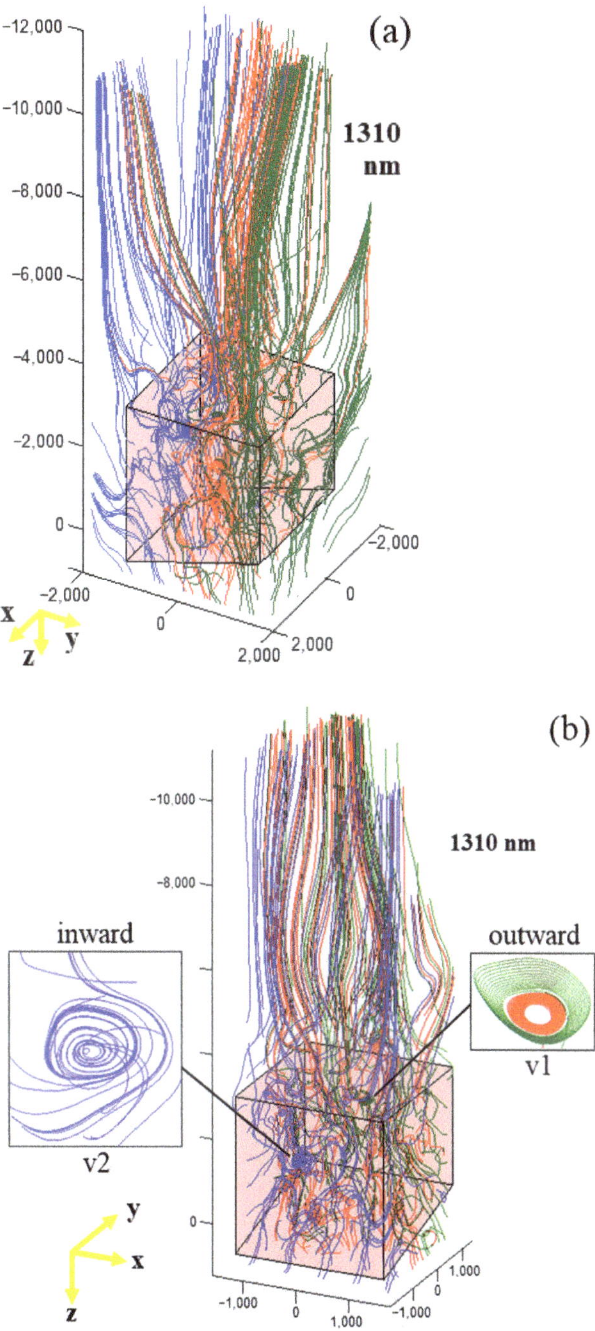

Figure 3. Overview (**a**) and the rotated view (**b**) of 3D Poynting vector distribution for the model of 1310 nm irradiation with the colour codes of streamlines: blue lines for long-base section (−*y* direction), red lines for middle section (around 0 position on *y* axis), and green lines for short-base section (+*y* direction). Insets in (**b**): Outward vortex v1 and Inward vortex v2.

The proposed switch enables the selection of the propagation routes for the optical signals at wavelengths of 1310 nm and 1550 nm, as shown in Figure 2, due to the opposite bending directions of photonic hooks. This phenomenon is unusual for two electromagnetic irradiations with such a small wavelength gap in the field of photonic hook study. This is mainly attributed to the separation of two high-magnitude areas of Poynting vectors as a1 and a2 shown in Figure 2c for the irradiation at the wavelength of 1310 nm. The corresponding contour seems to not perfectly match with the flow direction and density of Poynting vectors in 2D Poynting vector distribution (Figure 2c), especially for the gap between a1 and a2. Nevertheless, it can be properly explained by 3D Poynting vector analysis in this case. Here we use a 2D Poynting vector distribution of the median yz plane ($x = 0$) in greyscale as a screen to divide the 3D Poynting vector visualisation model for the 1310-nm-wavelength irradiation into $+x$ segment and $-x$ segment, as shown in the diagram in Figure 4a. View 1 ($+x$ segment visible) and view 2 ($-x$ segment visible) are defined as two observation angles to separately represent two segments of 3D Poynting vector distributions without allowing perspective in such a setup, as shown in Figure 4b,c. In these two figures, the profiles of the downward photonic hook and the high-magnitude areas, a1 and a2, are portrayed by black solid lines and yellow dashed lines, respectively. It is found that the profiles of a1 and a2 high-magnitude areas can be sequentially matched by the dense Poynting vector streamlines in the regions where are close to the median yz plane of $x = 0$ in $+x$ segment (view 1) and $-x$ segment (view 2), respectively. The gap between a1 and a2 shown in Figure 2c is attributed to this collective effect, and it leads to the unique downward photonic hook jointed by two high-magnitude areas for the irradiation at the wavelength of 1310 nm and spatially ensures optical isolation with the photonic hook (switching route) of 1550-nm-wavelength signal.

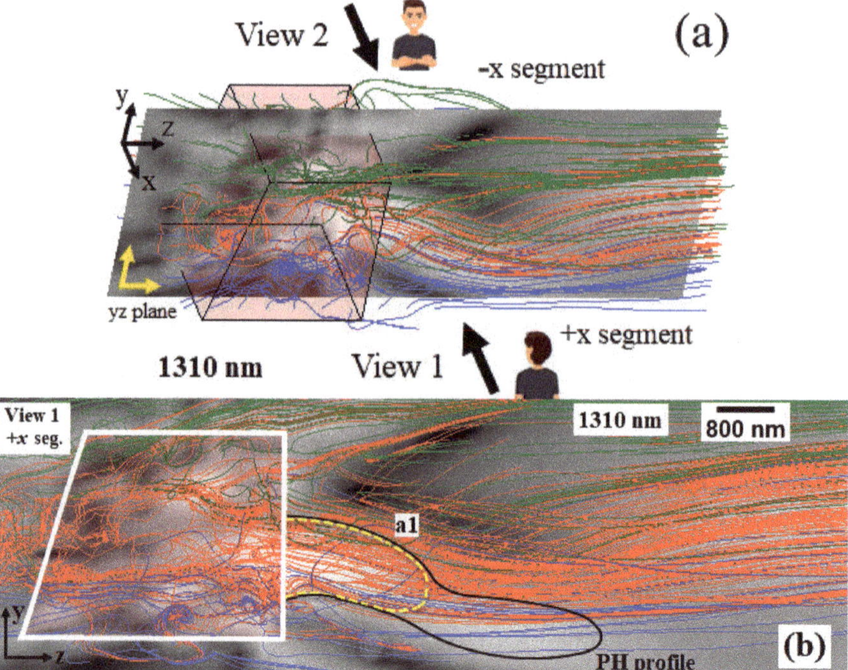

Figure 4. *Cont.*

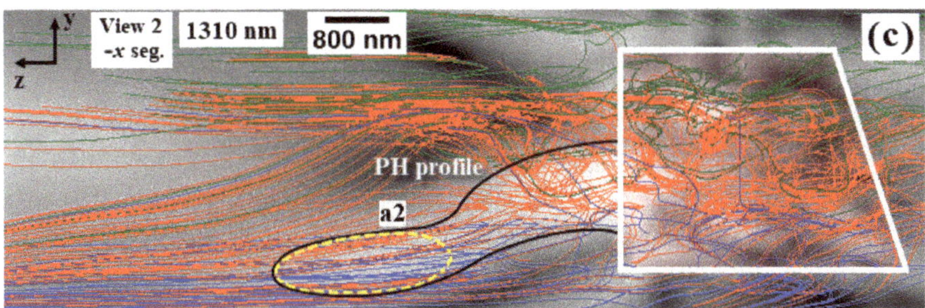

Figure 4. (**a**) Diagram of view 1 and view 2 segments and 3D Poynting vector distributions for (**b**) view 1 +x segment and (**c**) view 2 −x segment of the model with the irradiation at the wavelength of 1310 nm. Orientations of the particle in Figure 4b,c are reversed due to the change of frames of references for the observers in view 1 and view 2 as shown in Figure 4a.

Figure 5a,b illustrate the overview and close-up views of 3D Poynting vector distribution for the model of 1550-nm-wavelength irradiation. Compared to the distributions for the model of 1310-nm-wavelength irradiation shown in Figure 3, the Poynting vector streamlines in Figure 5 are less swirling in the trapezoid particle and more straight in the volume of the generated photonic hook, which reflects a smaller number of vortexes shown in Figure 5. Two key vortexes, v5 and v6, are indicated in Figure 5 as well. The photonic hook of the irradiation at a longer wavelength of 1550 nm appears in the region with the maximum density of Poynting vector streamlines, basically filling the central volume of the modelling space. These features are in accordance with the 2D Poynting distribution on the median yz plane for the same model shown in Figure 2d, which induces an almost straight, slightly upward photonic hook. Therefore, it is noted that the difference of spatial location of Poynting vector vortexes crucially influence the bending direction of the generated photonic hook and vary the subsequent propagation routes for the lights (optical signals) with similar wavelengths. This effect can be only represented by 3D Poynting vector distributions [32] and cause multiple localised high-magnitude regions of Poynting vectors to constitute a complete single photonic hook.

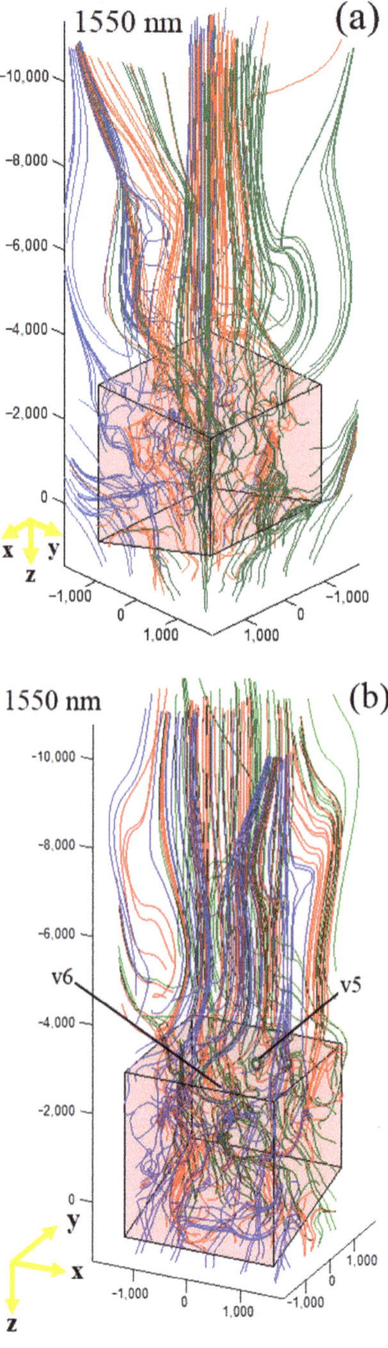

Figure 5. Overview (**a**) and the rotated view (**b**) of 3D Poynting vector distribution for the model of 1550 nm irradiation with the colour codes of streamlines: blue lines for long-base section (−*y* direction), red lines for middle section (around 0 position on y axis), and green lines for short-base section (+*y* direction).

3. Conclusions

The numerical simulation proves that the proposed near-field light-bending photonic switch as a Janus right-trapezoid-shaped dielectric particle embedded in the core facet of a planar waveguide can successfully switch and guide the optical signals at two common communication wavelengths, 1310 nm and 1550 nm, to different propagation routes based on photonic hook phenomenon. The opposite bending directions are found for the generated photonic hooks; the downward one for 1310-nm-wavelength irradiation is, unusually, separated into two high-magnitude areas by a Poynting vector flow. Three-dimensional Poynting vector analysis demonstrates that the corresponding high-magnitude areas can be sequentially matched by the Poynting vector streamlines in segments, which means that photonic hook can exist in a similar joining form. The corresponding mechanism is reported for the first time, and the future plan of the authors is to visualise similar 3D Poynting vector flows using virtual reality (VR) technology.

Author Contributions: Conceptualization, L.Y., O.V.M. and I.V.M.; methodology, L.Y.; software, L.Y., Z.W., and B.Y.; validation, Z.W. and O.V.M.; formal analysis, L.Y.; investigation, L.Y., O.V.M. and I.V.M.; resources, Y.X. and Y.E.G.; data curation, L.Y.; writing—original draft preparation, L.Y.; writing—review and editing, Z.W., O.V.M. and I.V.M.; visualization, L.Y.; supervision, L.Y., Z.W., O.V.M. and I.V.M.; project administration, L.Y., Z.W., O.V.M. and I.V.M.; funding acquisition, L.Y., Z.W., O.V.M. and I.V.M. All authors have read and agreed to the published version of the manuscript.

Funding: This research was funded by the Welsh European Funding Office (WEFO), European ERDF grants, grant number CPE 81400 and SPARCII c81133, the Royal Society International Exchanges Cost Share Scheme UK, grant number IEC\R2\202040, the Russian Foundation for Basic Research, grant number 21-57-10001, and the Tomsk Polytechnic University Development Program.

Institutional Review Board Statement: Not applicable.

Informed Consent Statement: Not applicable.

Data Availability Statement: This study did not report any data.

Conflicts of Interest: The authors declare no conflict of interest.

References

1. Jaeger, R.C. *Microelectronic Circuit Design*; McGraw-Hill: New York, NY, USA, 1997.
2. Rao, M.M. *Optical Communication*; Universities Press: Hyderabad, India, 2001.
3. Senior, J.M.; Jamro, M.Y. *Optical Fiber Communications*; Pearson/Prentice Hall: Hoboken, NJ, USA, 2009.
4. Jin, C.Y.; Wada, O. Photonic switching devices based on semiconductor nano-structures. *J. Phys. D Appl. Phys.* **2014**, *47*, 133001. [CrossRef]
5. Testa, F.; Pavesi, L. *Optical Switching in Next Generation Data Centers*; Springer: Beilin, Germany, 2017.
6. Bernstein, G.M.; Lee, Y.; Galver, A.; Martensson, J. Modeling WDM Wavelength Switching Systems for Use in GMPLS and Automated Path Computation. *IEEE/OSA J. Opt. Commun. Netw.* **2009**, *1*, 187–195. [CrossRef]
7. Cancellieri, G. *Single-Mode Optical Fibres*; Elsevier Science: Amsterdam, The Netherlands, 2014.
8. Ozcelik, D.; Parks, J.W.; Wall, T.A.; Stott, M.A.; Cai, H.; Parks, J.W.; Hawkins, A.R.; Schmidt, H. Optofluidic wavelength division multiplexing for single-virus detection. *Proc. Natl. Acad. Sci. USA* **2015**, *112*, 12933–12937. [CrossRef]
9. Sadot, D.; Boimovich, E. Tunable optical filters for dense WDM networks. *IEEE Commun. Mag.* **1998**, *36*, 50–55. [CrossRef]
10. Ramo, S.; Whinnery, J.R.; van Duzer, T. *Fields and Waves in Communications Electronics*; John Wiley and Sons: Hoboken, NJ, USA, 1984.
11. Prajzler, V.; Nekvindova, P.; Lyutakov, O. Flexible polymer planar optical waveguides. *Radioengineering* **2014**, *23*, 776–782.
12. Polky, J.N.; Mitchell, G.L. Metal-clad planar dielectric waveguide for integrated optics. *J. Opt. Soc. Am.* **1974**, *64*, 274–279. [CrossRef]
13. Minin, I.V.; Minin, O.V. *Diffractive Optics and Nanophotonics: Resolution below the Diffraction Limit*; Springer: Berlin, Germany, 2016.
14. Yue, L.; Minin, O.V.; Wang, Z.; Monks, J.N.; Snalin, A.S.; Minin, I.V. Photonic hook: A new curved light beam. *Opt. Lett.* **2018**, *43*, 771–774. [CrossRef] [PubMed]
15. Minin, I.V.; Minin, O.V.; Katyba, G.M.; Chernomyrdin, N.V.; Kurlov, V.N.; Zaytsev, K.I.; Yue, L.; Wang, Z.; Christodoulides, D.N. Experimental observation of a photonic hook. *Appl. Phys. Lett.* **2019**, *114*, 031105. [CrossRef]
16. Minin, I.V.; Minin, O.V.; Glinskiy, I.A.; Khalilubin, R.A.; Malureanu, R.; Lavrinenko, A.; Yakubovsky, D.I.; Volcow, V.S.; Ponomarev, D.S. Experimental verification of a plasmonic hook in a dielectric Janus particle. *Appl. Phys. Lett.* **2021**, *118*, 131107. [CrossRef]

17. Geints, Y.E.; Minin, O.V.; Yue, L.; Minin, I.V. Wavelength-Scale Photonic Space Switch Proof-of-Concept Based on Photonic Hook Effect. *Ann. Phys.* **2021**, *553*, 2100192. [CrossRef]
18. Soukoulis, C.M.; Wegener, M. Past achievements and future challenges in the development of three-dimensional photonic metamaterials. *Nat. Photonics* **2011**, *5*, 523–530. [CrossRef]
19. Yue, L.; Yan, B.; Monks, J.N.; Dhama, R.; Jiang, C.; Minin, O.V.; Minin, I.V.; Wang, Z. Full three-dimensional Poynting vector flow analysis of great field-intensity enhancement in specifically sized spherical-particles. *Sci. Rep.* **2019**, *9*, 20224. [CrossRef] [PubMed]
20. Yue, L.; Yan, B.; Wang, Z.B. Photonic nanojet of cylindrical metalens assembled by hexagonally arranged nanofibers for breaking the diffraction limit. *Opt. Lett.* **2016**, *41*, 1336–1339. [CrossRef] [PubMed]
21. Lee, S.Y.; Jeong, T.Y.; Jung, S.; Yee, K.J. Refractive Index Dispersion of Hexagonal Boron Nitride in the Visible and Near-Infrared. *Phys. Status Solidi B* **2018**, *256*, 1800417. [CrossRef]
22. Tan, C.Z. Determination of refractive index of silica glass for infrared wavelengths by IR spectroscopy. *J. Non-Cryst. Solids* **1998**, *223*, 158–163. [CrossRef]
23. MICROPOSIT®S1800®Photoresist Datasheet. Available online: https://amolf.nl/wp-content/uploads/2016/09/datasheets_S1 800.pdf (accessed on 25 January 2022).
24. Elson, J.M. Propagation in planar waveguides and the effects of wall roughness. *Opt. Express* **2001**, *9*, 461–475. [CrossRef]
25. Gomez-Correa, J.E.; Balderas-Mata, S.E.; Garza-Rivera, A.; Jaimes-Najera, A.; Trevino, J.P.; Coello, V.; Rogel-Salazar, J.; Chavez-Cerda, S. Mathematical-Physics of propagating modes in planar waveguides. *Rev. Mex. F'ısica E* **2019**, *65*, 218–222. [CrossRef]
26. Yan, B.; Yue, L.; Monks, J.N.; Yang, X.; Xiong, D.; Jiang, C.; Wang, Z. Superlensing plano-convex-microsphere (PCM) lens for direct laser nano-marking and beyond. *Opt. Lett.* **2020**, *45*, 1168–1171. [CrossRef]
27. Dietrich, P.I.; Blaicher, M.; Reuter, I.; Billah, M.; Hoose, T.; Hofmann, A.; Caer, C.; Dangel, R.; Offrein, B.; Troppenz, U.; et al. In situ 3D nanoprinting of free-form coupling elements for hybrid photonic integration. *Nat. Photon.* **2018**, *12*, 241–247. [CrossRef]
28. Stratton, J.A. *Electromagnetic Theory*, 1st ed.; McGraw-Hill: New York, NY, USA, 1941.
29. Bohren, C.F.; Huffman, D.R. *Absorption and Scattering of Light by Small Particles*; Wiley: Hoboken, NJ, USA, 1983.
30. Wang, Z.B.; Luk'yanchuk, B.S.; Hong, M.H.; Lin, Y.; Chong, T.C. Energy flow around a small particle investigated by classical Mie theory. *Phys. Rev. B* **2004**, *70*, 035418. [CrossRef]
31. Kumar, V.; Viswanathan, N.K. Topological structures in the Poynting vector field: An experimental realization. *Opt. Lett.* **2013**, *38*, 3886–3889. [CrossRef] [PubMed]
32. Minin, O.V.; Minin, I.V. Optical Phenomena in Mesoscale Dielectric Particles. *Photonics* **2021**, *8*, 591. [CrossRef]

Communication

A Closer Look at Photonic Nanojets in Reflection Mode: Control of Standing Wave Modulation

Ksenia A. Sergeeva [1], Alexander A. Sergeev [1], Oleg V. Minin [2] and Igor V. Minin [2,*]

1   Institute of Automation and Control Processes FEB RAS, 5 Radio Str., 690041 Vladivostok, Russia; kspetrovyh@mail.ru (K.A.S.); aleksandrsergeev@inbox.ru (A.A.S.)
2   Nondestructive School of Tomsk Polytechnic University, 30 Lenin Ave., 634050 Tomsk, Russia; oleg.minin@ngs.ru
*   Correspondence: prof.minin@gmail.com

Abstract: The photonic nanojet phenomenon is commonly used both to increase the resolution of optical microscopes and to trap nanoparticles. However, such photonic nanojets are not applicable to an entire class of objects. Here we present a new type of photonic nanojet in reflection mode with the possibility to control the modulation of the photonic nanojet by a standing wave. In contrast to the known kinds of reflective photonic nanojets, the reported one occurs when the aluminum oxide hemisphere is located at a certain distance from the substrate. Under illumination, the hemisphere generates a primary photonic nanojet directed to the substrate. After reflection, the primary nanojet acts as an illumination source for the hemisphere, leading to the formation of a new reflective photonic nanojet. We show that the distance between the hemisphere and substrate affects the phase of both incident and reflected radiation, and due to constructive interference, the modulation of the reflective photonic nanojet by a standing wave can be significantly reduced. The results obtained contribute to the understanding of the processes of photonic nanojet formation in reflection mode and open new pathways for designing functional optical devices.

Keywords: photonic nanojet; hemisphere; microstructure; interference; standing wave; reflection

Citation: Sergeeva, K.A.; Sergeev, A.A.; Minin, O.V.; Minin, I.V. A Closer Look at Photonic Nanojets in Reflection Mode: Control of Standing Wave Modulation. *Photonics* **2021**, *8*, 54. https://doi.org/10.3390/photonics8020054

Received: 27 December 2020
Accepted: 15 February 2021
Published: 17 February 2021

Publisher's Note: MDPI stays neutral with regard to jurisdictional claims in published maps and institutional affiliations.

1. Introduction

Along with plasmonics, dielectric photonics [1] is associated with the processes of localization of light and is aimed at increasing the efficiency of the interaction of incident radiation with matter. The progress in this direction contributes to the development of compact optical radiation control devices, opening opportunities for the creation of new sensing structures [2], detection of nano- and microparticles [3–6], nanoparticle manipulation [7,8], luminescence [9,10] and Raman scattering enhancement [11], and both optical [12,13] and terahertz [14] super-resolution microscopy. Moreover, dielectric photonic structures are free from the disadvantages associated with extremely high losses that occur with plasmonic structures, which significantly increases the efficiency of devices based on them [15].

Localization of light requires sharp focusing, which can be achieved by high-numerical-aperture diffractive optical elements [16]. To date, the most popular approach for micro- and nanoobject manipulation is the use of various metalenses. In general, a metalens consists of many distributed microstructures that adjust the incident light to be focused from several tens of micrometers [17,18] to several millimeters [19]. For example, some authors [20] used a reflective metalens with a size of 100 μm × 100 μm, which corresponds to lens dimensions of 93.98λ × 93.98λ at λ = 1064 nm with a focal length of 108λ, almost equal to the lens diameter. Moreover, the beam waist radius was equal to 0.62λ, i.e., greater than the diffraction limit. These are the typical size and parameters of microlenses [16].

Although the light cannot be directly focused into the subwave region, for certain sizes of the dielectric microparticle, it can be "limited" in all three dimensions due to the specific interference of the radiation transmitted and scattered on the microstructure—an effect called a "photonic nanojet" (PNJ) [13,21]. In the photonic nanojet phenomenon, the scattering is characterized by the refractive index n as well as the Mie size parameter $q = 2\pi Rn/\lambda$, if the dielectric particle has no dissipation. Usually, to form a PNJ, the size parameter is q = (10 ... 20)π [22,23] and takes an intermediate value between the dipole approximation for smaller particles (R $\ll$ λ) and the geometrical optics approximation for the largest particles (including lenses, q > 100π). In contrast to metalenses, the use of a photonic nanojet allows for localizing the incident radiation into the subwavelength region using a mesoscale particle-lens near its surface, with a characteristic size on the order of or slightly larger than the wavelength with beam waist radius less than the diffraction limit [21–23]. Accordingly, the size of the focusing structure can be reduced by about 100 times, compared to classical high-numerical-aperture lenses with increasing focused field localization below the diffraction limit. The size of a single phase-tuning element in metalenses has a wavelength scale, which is comparable with the size of a PNJ-generating microstructure [21]. In this regard, the same technique used for metalens fabrication could be used for the creation of arrays of PNJ-generated microstructures, opening a way for the development of new functional structures.

Subwavelength light localization via photonic nanojets not only simplifies the trapping of nano- and micro-objects but also allows us to reduce the intensity of "trapping light". In [24], a PNJ with localization intensity (ratio of the intensities of the focal spot and incident light) equal to 4 was used to trap 50-nm-sized gold nanoparticles and track the trapping event through backscattered light. The minimal PNJ intensity for nanoparticle trapping seems to be around 3 [7], allowing operation with sub-100-nm-sized particles. Enhancement of the photoluminescence (PL) or Raman scattering signal is the other promising field for PNJ application. A tenfold enhancement of ZnO thin-film PL was achieved using a PNJ generated by 5-µm-sized fused silica and polystyrene microspheres [25]. This PL enhancement arises from modifications of the optical density of states in ZnO due to the Purcell effect [26] with a threshold starting from a PNJ intensity of 3.5. In turn, the lowest PNJ intensity for Raman scattering enhancement was reported to be 2.3 [27].

Most studies of photonic nanojets have aimed at obtaining sharp-focused high-intensity regions of radiation localization in transmission mode, so such functional structures are not suitable for "traditional" optical and fluorescent microscopes operated in reflection geometry. In this regard, the development of new functional structures generating PNJs in reflection mode allows us to combine the unique properties of PNJs and simplify techniques for their implementation. In general, the formation of a reflective photonic nanojet from a dielectric microparticle occurs due to the interaction of reflected light, passed through the parental microparticle, with the incident light, leading to modulation of the photonic nanojet by a standing wave [28–31]. This feature has great practical importance since it allows for achieving a relatively high localization intensity in the simplest experimental geometry. The process of PNJ formation in this geometry is well-studied theoretically [28–32], and one can find methods for its implementation in real experiments. For example, the stability of the Brownian motion of nanoparticles could be improved by using a reflective photonic nanojet [16,30]. It is also well known that optical devices, like a microscope, or surface-enhanced Raman scattering and photoluminescence-enhancing structures, can operate in both transmission and reflection modes [33–35].

Here we show that it is possible to avoid modulation of the photonic nanojet by a standing wave by placing, for example, an aluminum oxide hemisphere [36] at a certain distance (a gap) from a silicon substrate and using a low-coherence excitation source. The results obtained indicate that the generated PNJ shows behavior similar to that of a "classical" PNJ in reflection mode, so it is possible to distinguish the processes involved in its formation. To provide a link between numerical and experimental studies of the formation of this kind of PNJ, we fabricated an ordered array of aluminum oxide microspheres with numerically predicted optical and size parameters and performed direct experimental measurements of the reflective photonic nanojets generated. This, in turn, opens the way for microminiaturization of photonics for various purposes and applications.

2. Materials and Methods

The numerical simulations were performed via finite-difference time-domain modeling [37], performed in Matlab software. The simulation domain was a triangular mesh with a perfectly matched layer as a boundary. The mesh size was set equal to $\lambda/25$, where λ = 532 nm is an illuminating wavelength, to assure balance between accuracy and calculated speed. Space and time derivatives were both calculated via second-order accurate centered difference. The time step was determined by the size of the triangular mesh and selected to ensure numerical stability. A plane wave TE-polarized light with unit intensity was generated on the left side of the simulation domain. Calculations were conducted for two positions of a plane wave source with unit intensity: (i) at 3 μm above the hemisphere, which was a compromise between model accuracy and calculation time, and (ii) at 100 nm above the parental microstructure to exclude parasitic interference between the light source and the PNJ generated in reflection mode. To provide a link with further practical applications of the PNJ studied, the refractive index of the propagating media was chosen to be n_c = 1.545, which is common for various polymeric materials [38]. The refractive indices of polymethylmethacrylate (PMMA), alumina, and silicon are 1.495, 1.7731, and 4.22469, respectively, at the incident wavelength [39,40].

Direct experimental studies of the photonic nanojet formation were carried out via a specially built setup (Figure 1). This setup allows one to simultaneously record the luminescent signal from an area on the order of three square micrometers. A charge-coupled device (CCD) array and an Andor high-speed spectrometer were used for image capture and spectral measurements, correspondingly. The size of the CCD was 35.9 × 24 mm with a pixel size of 5.9 μm, and the resolution was 6016 × 4016 pixels. The Andor spectrometer was equipped with a Shamrock 750 monochromator and an iStar camera. The monochromator had three motorized diffraction gratings with 600, 1200, and 2400 lines per millimeter, and the maximum spectral resolution was 0.3 nm. The iStar camera is a cooled CCD with a pixel size of 13 × 13 microns, providing a maximum shooting speed of 1.3 ns. A Nikon λ Plan Apo series microscope lens (NA0.9, 100x) was used to construct images. The lens was mounted on a piezoelectric positioner (movement range along the Z-axis, 400 μm; accuracy, 10 nm; minimal step, 4 nm). An adjustable neutral optical density filter (variable optical density from 0.1 to 4) was used to balance the incident radiation intensity for the constructed image and spectral intensity measurements. During image construction, the objective formed a slightly divergent light beam, which was then divided into two channels. The first beam was directed to the CCD that captured the image, and the second one was collected by the fiber input of the spectrometer (core diameter, 300 μm), acting as a pinhole in the confocal scheme. The CCD and the fiber input of the spectrometer were placed at the same distance from the beam-splitting cube; hence, the images of the working area captured by CCD and the one directed to the spectrometer had the same lateral dimensions.

The NKT Photonics SuperK Extreme broadband coherent light source was used as an incident light source; it generates radiation in the range of 450–1800 nm and was equipped with an acousto-optic filter unit with the possibility to select the required wavelength with a spectral half-width of 5 nm and average power of 3 mW. This source has low spatial coherence and emits an almost unpolarized light, which allows one to reduce as much as possible the occurrence of interference events during measurements.

The fabrication of dielectric structures with precalculated geometric parameters was carried out by a combination of methods of electron beam lithography, vacuum deposition, and selective chemical etching. These methods are the standard technological approaches used in silicon microelectronics and, thus, could be used in further serial production of functional optical devices based on PNJ phenomena. A schematic representation of the dielectric microstructure fabrication process is shown in Figure 2a and, in general, consisted of three steps. In the first step, a PMMA layer was spin-coated onto a silicon substrate. To provide higher mechanical stability of the PMMA film and avoid probable damage during further development, the film was annealed at 150 °C for 3 h, resulting in total removal of the remaining solvent molecules and reduction of the internal stresses in the film due to its transformation from liquid to solid state. After that, the pattern for the future microparticle array with the required hemispherical profile was formed in the PMMA film via electron beam lithography. The exposure dose and scanning rate were adjusted in such a way as to ensure a hemispherical profile to be written in the volume of PMMA film, keeping the required gap distance between the microstructures and substrate. It is well known that PMMA is a positive resist, and its molecules break during exposure to an electron beam. As a result, the exposed areas become soluble in the developer solution (9:1 mixture of isopropyl alcohol with water). The resulting array of holes in the PMMA film acted as a template for microparticle creation via the vacuum deposition technique. The quality of the template, e.g., the depth of the microholes and its surface profile, was controlled via the atomic-force microscopy (AFM) technique (Figure 2b), which revealed that the average upper diameter of microholes was about 1000 ± 40 nm, with 500 ± 10 nm depth. The alumina deposition was performed using a PVD-2EB2R11 vacuum deposition device (ADVAVAC, Canada) equipped with a magnetron sputtering system, allowing deposited layer thickness control of up to 1 nm. To increase adhesion, before deposition, the sample was activated by argon plasma.

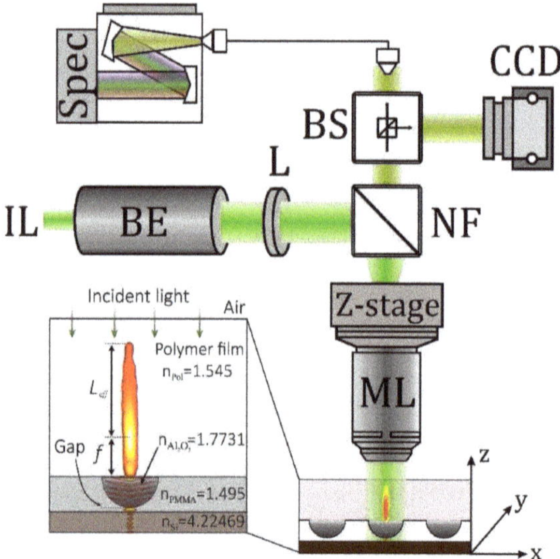

Figure 1. Specially built setup for photonic nanojet studies: IL, incident light; BE, beam expander; L, collimating lens; NF, Notch filter; ML, microscope lens; Z-stage, piezoelectric positioner; BS, beamsplitter; Spec, spectrometer; CCD, camera. The inset shows the material definitions.

In general, the fabricated arrays occupy an area of 400×400 μm. In the case of a 1-μm-diameter aluminum oxide hemisphere, the array contains 17.5 thousand single microstructures. Figure 2c shows a scanning electron microscope (SEM) image of a 135×100 μm area with more than 2000 uniformly distributed dielectric microstructures of the same lateral size. In turn, a 0.9NA-100× microscope lens, used in our setup for image construction, provided simultaneous capture of the PNJs from approximately 1000 microstructures (Figure 2d), giving plenty of room for precise high-resolution optical measurements. It should be noted that microstructures could be resolved using even a 0.6NA microscope lens with 40× or 60× magnification. However, a low-resolution lens does not allow direct measurements of a single PNJ's intensity, because there are about 20 microstructures in a field of view of the fiber-optical signal collection system.

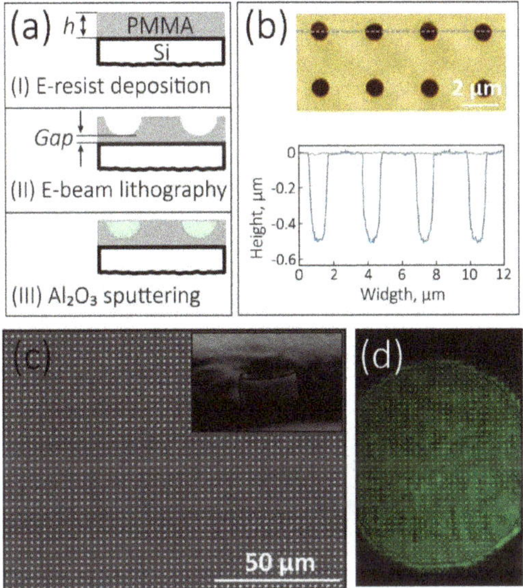

Figure 2. (**a**) A schematic illustration of the fabrication process of aluminum oxide hemispheres; (**b**) AFM image of the polymethylmethacrylate (PMMA) surface after production of the hemisphere templates (scale bar is 2 µm), and the corresponding depth profiles before (blue curve) and after (gray curve) aluminum oxide deposition; (**c**) SEM image top-view of the aluminum oxide hemisphere array (scale bar is 50 µm); the inset shows a cross section of PMMA film cut through the microstructure array, indicating their hemispherical shape; (**d**) an array of photonic nanojets generated in reflection mode, captured with a 0.9NA microscope lens.

3. Results and Discussion

The numerically simulated pattern of the reflective photonic nanojet's modulation by a standing wave is shown in Figure 3a. It was found that the use of a low-coherence light source made it possible to reduce the interference between the incident and reflected light, resulting in "uniform" PNJ formation, without the appearance of the interference pattern (Figure 3b).

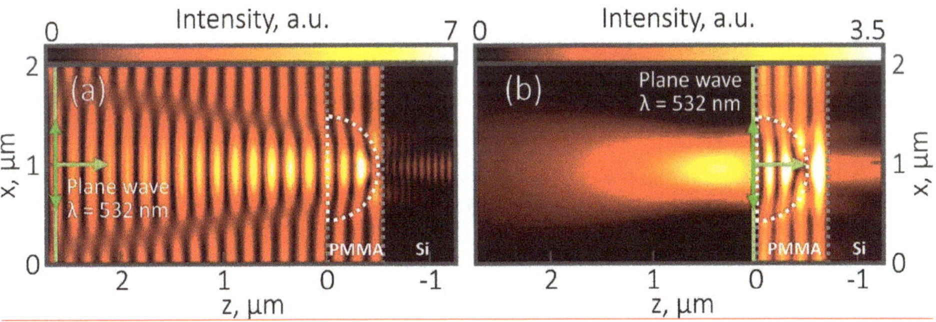

Figure 3. Numerical simulation of photonic nanojet (PNJ) formation in reflection mode: (**a**) the intensity distribution for a PNJ formed via modulation by a standing wave (for calculations, the light source was placed at 3 µm above the hemisphere); (**b**) the intensity distribution for a PNJ formed without modulation (for calculations, the light source was placed at 100 nm above the hemisphere). Incident light (λ = 532 nm in the air) propagates from left to right. For clarity, the hemisphere position is marked by white dashed lines. The calculated intensity distributions were normalized to the incident light intensity and plotted up to the maximal PNJ intensity.

It is interesting to note that both types of PNJs (generated by or without interference with incident light) have pretty similar properties: (i) the maximum localization intensity is separated from the hemisphere's surface and located at $\approx 0.8\lambda$ from it; (ii) the PNJ length (L_{eff}), defined as the distance where the maximal PNJ intensity (I_{max}) decreases by e times, is also almost the same and equal to 1.7 µm or $\approx 3\lambda$ (in a case of an "interference-based" PNJ, this distance was estimated for interference maxima); (iii) the FWHM of PNJs in a maximum-intensity area is also similar and equal to 0.75 λ. The quality factor (Q) was determined as [22,41]

$$Q = \frac{I_{max} \cdot L_{eff}}{FWHM},$$ (1)

where FWHM is the full width at half-maximum of the PNJ intensity; for PNJs generated without modulation by a standing wave, this is equal to 7.37 and close to the square root of Q for a PNJ with modulation (46.19), which also indicates the connection between these two PNJ types.

Taking into account the value of the quality factor and light intensity in the localization area, we believe that the new type of reflective photonic nanojet described above may replace different types of objective lenses (metalens) or arrays of lenses [42,43] in different miniaturized environments, including lab-on-a-chip devices, and could be suitable for the manipulation of nanoparticles and the enhancement of PL and Raman scattering signals as well. However, the use of such devices and applications is a challenge for future research.

To provide experimental verification, we obtained the PNJ intensity distribution across its propagation length using the synchronizing spectrometer with the piezoelectric positioner to collect the intensity at every movement step (≈ 4 nm). The experimentally measured PNJ intensity distribution was in good agreement with the numerically calculated one (Figure 4). Moreover, the PNJ images captured at the maximum-intensity point (0.4 µm on the Z-axis), on the end of an effective PNJ length (1.5 µm on the Z-axis), and at the very end of the PNJ, where its intensity becomes close to the background level (2.8 µm on the Z-axis), were also in a good agreement with the calculated ones. The results obtained prove the possibility of PNJ generation in reflection mode without modulation by a standing wave.

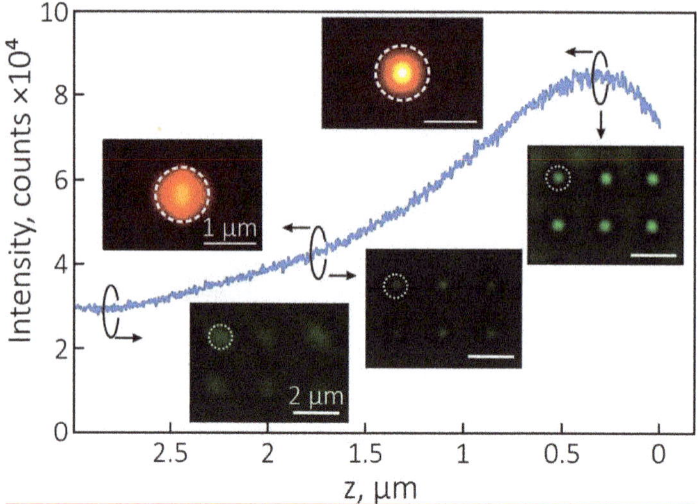

Figure 4. Intensity scan across the Z-axis on the incident wavelength. Top insets illustrate the calculated PNJ intensity distributions at the marked points (scale bar is 1 µm). Bottom insets show the captured images of PNJs at the marked points (scale bar is 2 µm). The zero point on the Z-scale corresponds to the surface of the hemisphere.

The next question to be solved is to distinguish the process of PNJ formation in reflection mode due to interference between reflected and incident light. The intensity distributions for both PNJ types (Figure 5a) illustrate the similar behavior of the PNJs, mentioned above in the Introduction section. This allows one to conclude that the interaction of a bare reflected PNJ with incident light can be described by the well-known interference equation [44] as follows:

$$I = I_{PNJ} + I_0 + 2 \cdot \sqrt{I_{PNJ} \cdot I_0} \cdot \cos(2\pi k), \tag{2}$$

where I_{PNJ} is the intensity of the PNJ due to reflection from the substrate, and I_0 is the incident light intensity; $k = (n_{hs} \cdot R_{hs} + I\varphi) \cdot L_Z$ is the reflected light phase shift caused by its passing through the gap ($I\varphi$) and the hemisphere with refractive index n_{hs} and radius R_{hs}, as well, during its backward propagation along the Z-axis over the distance L_Z. It should be noted that inside the gap the additional interference pattern occurs, which could either enhance or quench the light reflected from the substrate. This feature is likely a reason for the nonmonotonic change in PNJ intensity with increasing the gap [36].

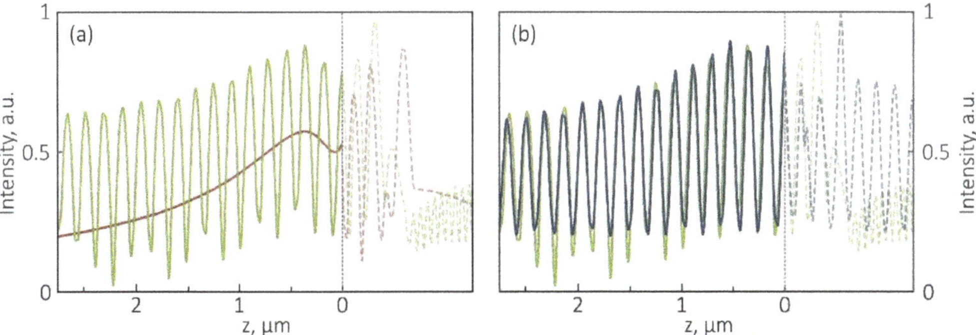

Figure 5. (**a**) Intensity distribution profiles for PNJs generated in reflection mode due to interference between reflected and incident light (green curve) and only due to reflected light (red line). (**b**) A comparison of interference patterns for PNJs formed by the interaction of reflected and incident light (green line, the same as in Figure 5a) and the calculated interference pattern between reflected light (which is the red curve on Figure 5a) and incident light with unity intensity (blue curve). The surface of the hemisphere is marked by a grey dashed line, while the light traces inside the hemisphere and those passed through it to the substrate are shown by semitransparent curves.

As a result, the "simulated" interference patterns [45] are quite well matched with the numerically calculated intensity distribution for PNJs formed due to interference between reflected and incident light (Figure 5b). Some mismatch in the modulation depth at distances above the effective PNJ length, and a significant mismatch in interference patterns in the gap and substrate area, may indicate the presence of additional factors affecting PNJ formation in reflection mode.

Analysis of the data presented in Figures 3–5 allows us to assume the following mechanism of PNJ formation due to only light reflected from the substrate for the case when there is a certain gap between the parental microstructure and the substrate. After illumination, the hemisphere forms a primary PNJ on its shadow surface (this is a "classical" PNJ in transmission mode) [46]. This primary PNJ is located slightly apart ($\approx$100 nm or $\frac{1}{2}$ gap) from the hemisphere's surface. When the primary PNJ is reflected back from the substrate, it acts as a new (effective) illumination source for the hemisphere, leading to the formation of a secondary PNJ, which appears to be the one PNJ formed due to bare reflection from a substrate. In fact, by choosing the value of the gap, we controlled the phases of radiation both incident and reflected from the substrate, and due to constructive interference, the modulation of the reflective photonic nanojet by a standing wave was significantly reduced. From this point of view, concerning the new excitation source position, the secondary PNJ is formed as in transmission mode and, thus, could exist without modulation by the standing wave.

4. Conclusions

A new type of photonic nanojet in reflection mode without modulation by a standing wave was numerically and experimentally predicted herein. This nanojet was generated by an aluminum oxide micro-hemisphere placed at a certain gap from the silicon substrate and occurred under low-coherence light illumination. In this case, the parental microstructure first generated a PNJ in transmission mode that, after reflection from the substrate, acted as a new effective excitation source for the microstructure. The exact gap value depended on the parental microstructure parameters: size, shape, refractive index contrast with surrounding media, and substrate parameters (i.e., reflection coefficient, refractive index, etc. [21,28,47]) that determine the properties of light reflection from the substrate, thus affecting the final PNJ properties. Taking into account that the PNJ properties depend on refractive index contrast [21,27,39,48], it is possible to adjust these properties by changing the material of the microstructure, its surrounding media, substrate parameters, and/or the gap value. The results obtained are promising for optical sensing, surface-enhanced Raman scattering [11,27], optical data processing, nanoparticle manipulations, sorting and imaging of micro- and nanoobjects, and studies in reflection geometry using a regular confocal microscope [49].

Author Contributions: Conceptualization, I.V.M. and O.V.M.; investigation K.A.S. and A.A.S.; writing—original draft preparation, K.A.S. and A.A.S.; writing—review and editing, I.V.M. and O.V.M. All authors have read and agreed to the published version of the manuscript.

Funding: The experimental studies of photonic nanojet formation in reflection mode were supported by the Russian Science Foundation (project 20-72-00066).

Data Availability Statement: Experimental data is available from the corresponding author upon reasonable request.

Acknowledgments: I.V.M. and O.V.M. acknowledge the Tomsk Polytechnic University Competitiveness Enhancement Program.

Conflicts of Interest: The authors declare no conflict of interest.

References

1. Baranov, D.; Zuev, D.; Lepeshov, S.; Kotov, O.; Krasnok, A.; Evlyukhin, A.; Chichkov, B. All-dielectric nanophotonics: The quest for better materials and fabrication techniques. *Optica* **2017**, *4*, 814–825. [CrossRef]
2. Sergeeva, K.A.; Tutov, M.V.; Voznesenskiy, S.S.; Shamich, N.I.; Mironenko, A.Y.; Sergeev, A.A. Highly-sensitive fluorescent detection of chemical compounds via photonic nanojet excitation. *Sens. Act. B* **2020**, *305*, 127354. [CrossRef]
3. Li, Y.; Xin, H.; Liu, X.; Zhang, Y.; Lei, H.; Li, B. Trapping and detection of nanoparticles and cells using a parallel photonic nanojet array. *ACS Nano* **2016**, *10*, 5800–5808. [CrossRef]
4. Ren, Y.X.; Zeng, X.; Zhou, L.M.; Kong, C.; Mao, H.; Qiu, C.W.; Tsia, K.K.; Wong, K.K. Photonic nanojet mediated backaction of dielectric microparticles. *ACS Photonics* **2020**, *7*, 1483–1490. [CrossRef]
5. Tomitaka, A.; Arami, H.; Ahmadivand, A.; Pala, N.; McGoron, A.J.; Takemura, Y.; Febo, M.; Nair, M. Magneto-plasmonic nanostars for image-guided and NIR-triggered drug delivery. *Sci. Rep.* **2020**, *10*, 1–10. [CrossRef] [PubMed]
6. Yu, X.; Li, A.; Zhao, C.; Yang, K.; Chen, X.; Li, W. Ultrasmall semimetal nanoparticles of bismuth for dual-modal computed tomography/photoacoustic imaging and synergistic thermoradiotherapy. *ACS Nano* **2017**, *11*, 3990–4001. [CrossRef]
7. Li, Y.C.; Xin, H.B.; Lei, H.X.; Liu, L.L.; Li, Y.Z.; Zhang, Y.; Li, B.J. Manipulation and detection of single nanoparticles and biomolecules by a photonic nanojet. *Light Sci. Appl.* **2016**, *5*, e16176. [CrossRef]
8. Minin, I.V.; Minin, O.V.; Cao, Y.; Liu, Z.; Geints, Y.E.; Karabchevsky, A. Optical vacuum cleaner by optomechanical manipulation of nanoparticles using nanostructured mesoscale dielectric cuboid. *Sci. Rep.* **2019**, *9*, 12748. [CrossRef] [PubMed]
9. Zhang, W.; Lei, H. Fluorescence enhancement based on cooperative effects of a photonic nanojet and plasmon resonance. *Nanoscale* **2020**, *12*, 6596–6602. [CrossRef]
10. Cai, Y.Y.; Collins, S.S.; Gallagher, M.J.; Bhattacharjee, U.; Zhang, R.; Chow, T.H.; Ahmadivand, A.; Ostovar, B.; Al-Zubeidi, A.; Wang, J.; et al. Single-Particle Emission Spectroscopy Resolves d-Hole Relaxation in Copper Nanocubes. *ACS Energy Lett.* **2019**, *4*, 2458–2465. [CrossRef]

11. Das, G.M.; Laha, R.; Dantham, V.R. Photonic nanojet-mediated SERS technique for enhancing the Raman scattering of a few molecules. *J. Raman Spectrosc.* **2016**, *47*, 895–900. [CrossRef]

12. Wang, Z.; Guo, W.; Li, L.; Luk'yanchuk, B.; Khan, A.; Liu, Z.; Chen, Z.; Hong, M. Optical virtual imaging at 50 nm lateral resolution with a white-light nanoscope. *Nat. Commun.* **2011**, *2*, 218. [CrossRef] [PubMed]

13. Yang, H.; Trouillon, R.; Huszka, G.; Gijs, M.A. Super-resolution imaging of a dielectric microsphere is governed by the waist of its photonic nanojet. *Nano Lett.* **2016**, *16*, 4862–4870. [CrossRef] [PubMed]

14. Ahi, K.; Jessurun, N.; Hosseini, M.-P.; Asadizanjani, N. Survey of terahertz photonics and biophotonics. *Opt. Eng.* **2020**, *59*, 061629. [CrossRef]

15. Ostovar, B.; Cai, Y.Y.; Tauzin, L.J.; Lee, S.A.; Ahmadivand, A.; Zhang, R.; Nordlander, P.; Link, S. Increased intraband transitions in smaller gold nanorods enhance light emission. *ACS Nano* **2020**, *14*, 15757–15765. [CrossRef] [PubMed]

16. Dholakia, K.; Čižmár, T. Shaping the future of manipulation. *Nat. Photon.* **2011**, *5*, 335. [CrossRef]

17. Li, Z.; Liu, W.; Li, Z.; Tang, C.; Cheng, H.; Li, J.; Chen, X.; Chen, S.; Tian, J. Nonlinear metasurfaces: Tripling the capacity of optical vortices by nonlinear metasurface. *Laser Phot. Rev.* **2018**, *12*, 1870049. [CrossRef]

18. Arbabi, E.; Arbabi, A.; Kamali, S.M.; Horie, Y.; Faraji-Dana, M.; Faraon, A. MEMS-tunable dielectric metasurface lens. *Nat. Comm.* **2018**, *9*, 812. [CrossRef] [PubMed]

19. Yin, X.; Steinle, T.; Huang, L.; Taubner, T.; Wuttig, M.; Zentgraf, T.; Giessen, H. Beam switching and bifocal zoom lensing using active plasmonic metasurfaces. *Light Sci. Appl.* **2017**, *6*, e17016. [CrossRef]

20. Tkachenko, G.; Stellinga, D.; Ruskuc, A.; Chen, M.; Dholakia, K.; Krauss, T.F. Optical trapping with planar silicon metalenses. *Opt. Lett.* **2018**, *43*, 3224–3227. [CrossRef]

21. Zhu, J.; Goddard, L.L. All-dielectric concentration of electromagnetic fields at the nanoscale: The role of photonic nanojets. *Nanoscale Adv.* **2019**, *1*, 4615–4643. [CrossRef]

22. Heifetz, A.; Kong, S.-C.; Sahakian, A.V.; Taflove, A.; Backman, V. Photonic Nanojets. *J. Comput. Theor. Nanosci.* **2009**, *6*, 1979–1992. [CrossRef]

23. Littlefield, A.J.; Zhu, J.; Messinger, J.F.; Goddard, L.L. Photonic Nanojets. *Opt. Photonics News* **2020**, *39*, 2.

24. Yang, H.; Cornaglia, M.; Gijs, M.A. Photonic nanojet array for fast detection of single nanoparticles in a flow. *Nano Lett.* **2015**, *15*, 1730–1735. [CrossRef] [PubMed]

25. Yan, Y.; Zeng, Y.; Wu, Y.; Zhao, Y.; Ji, L.; Jiang, Y.; Li, L. Ten-fold enhancement of ZnO thin film ultraviolet-luminescence by dielectric microsphere arrays. *Opt. Exp.* **2014**, *22*, 23552–23564. [CrossRef] [PubMed]

26. Artemyev, M.V.; Woggon, U.; Wannemacher, R.; Jaschinski, H.; Langbein, W. Light trapped in a photonic dot: Microspheres act as a cavity for quantum dot emission. *Nano Lett.* **2001**, *1*, 309–314. [CrossRef]

27. Patel, H.; Kushwaha, P.; Swami, M. Photonic nanojet assisted enhancement of Raman signal: Effect of refractive index contrast. *J. Appl. Phys.* **2018**, *123*, 023102. [CrossRef]

28. Yue, L.; Yan, B.; Monks, J.N.; Dhama, R.; Wang, Z.; Minin, O.V.; Minin, I.V. Photonic jet by a near-unity-refractive-index sphere on a dielectric substrate with high index contras. *Ann. Phys.* **2018**, *530*, 1800032. [CrossRef]

29. Minin, I.V.; Minin, O.V.; Pacheco-Peña, V.; Beruete, M. Subwavelength, standing-wave optical trap based on photonic jets. *Quantum Electron.* **2016**, *46*, 555–557. [CrossRef]

30. Minin, O.V.; Geints, Y.E.; Zemlyanov, A.A.; Minin, O.V. Specular-reflection photonic nanojet: Physical basis and optical trapping application. *Opt. Exp.* **2020**, *28*, 22690. [CrossRef]

31. Minin, I.V.; Liu, C.Y.; Yang, Y.C.; Staliunas, K.; Minin, O.V. Experimental observation of flat focusing mirror based on photonic jet effect. *Sci. Rep.* **2020**, *10*, 8459. [CrossRef]

32. Liu, C.Y.; Chung, H.J.; Hsuan-Pei, E. Reflective photonic hook achieved by a dielectric-coated concave hemicylindrical mirror. *Josa B* **2020**, *37*, 2528–2533. [CrossRef]

33. Wen, Y.; Yu, H.; Zhao, W.; Li, P.; Wang, F.; Ge, Z.; Wang, X.; Liu, L.; Li, W.J. Scanning Super-Resolution Imaging in Enclosed Environment by Laser Tweezer Controlled Superlens. *Biophys. J.* **2020**, *119*, 2451–2460. [CrossRef] [PubMed]

34. Biener, G.; Greenbaum, A.; Isikman, S.O.; Lee, K.; Tseng, D.; Ozcan, A. Combined reflection and transmission microscope for telemedicine applications in field settings. *Lab Chip.* **2011**, *11*, 2738–2743. [CrossRef]

35. Lee, M.; Yaglidere, O.; Ozcan, A. Field-portable reflection and transmission microscopy based on lensless holography. *Biomed. Opt. Exp.* **2011**, *2*, 2721. [CrossRef]

36. Sergeev, A.A.; Sergeeva, K.A. Functional dielectric microstructure for photonic nanojet generation in reflection mode. *Opt. Mat.* **2020**, *110*, 110503. [CrossRef]

37. Taflove, A.; Hagness, S. *Computational Electrodynamics: The Finite Difference Time Domain Method*; Artech House: Norwood, MA, USA, 1998.

38. Wiederseiner, S.; Andreini, N.; Epely-Chauvin, G.; Ancey, C. Refractive-index and density matching in concentrated particle suspensions: A review. *Exp. Fluids* **2011**, *50*, 1183–1206. [CrossRef]

39. Loste, J.; Lopez-Cuesta, J.-M.; Billon, L.; Garay, H.; Save, M. Transparent polymer nanocomposites: An overview on their synthesis and advanced properties. *Prog. Polym. Sci.* **2019**, *89*, 133–158. [CrossRef]

40. Shaker, L.M.; Al-Amiery, A.A.; Amir, A.; Kadhum, H.; Takri, M.S. Manufacture of Contact Lens of Nanoparticle-Doped Polymer Complemented with ZEMAX. *Nanomaterials* **2020**, *10*, 2028. [CrossRef]

41. Born, M.; Wolf, E. *Principles of Optics: Electromagnetic Theory of Propagation, Interference and Diffraction of Light*; Cambridge University Press: Cambridge, UK, 1999; pp. 286–411.
42. Luo, H.; Yu, H.; Wen, Y.; Zhang, T.; Li, P.; Wang, F.; Liu, L. Enhanced high-quality super-resolution imaging in air using microsphere lens groups. *Opt. Lett.* **2020**, *45*, 2981–2984. [CrossRef]
43. Zhang, T.; Yu, H.; Li, P.; Wang, X.; Wang, F.; Shi, J.; Liu, Z.; Yu, P.; Yang, W.; Wang, Y.; et al. Microsphere-based super-resolution imaging for visualized nanomanipulation. *ACS Appl. Mater. Interfaces* **2020**, *12*, 48093–48100. [CrossRef] [PubMed]
44. Espinoza, F. Interference and Standing Waves. In *Wave Motion as Inquiry*; Springer: Berlin, Germany, 2017.
45. Zemánek, P.; Jonáš, A.; Šrámek, L.; Liška, M. Optical trapping of nanoparticles and microparticles by a Gaussian standing wave. *Opt. Lett.* **1999**, *24*, 1448–1450. [CrossRef]
46. Wen, Y.; Yu, H.; Zhao, W.; Wang, F.; Wang, X.; Liu, L.; Li, W.J. Photonic Nanojet Sub-Diffraction Nano-Fabrication with in situ Super-Resolution Imaging. *IEEE Trans. Nanotechnol.* **2019**, *18*, 226–233. [CrossRef]
47. Zhang, X.A.; Chen, I.-T.; Chang, C.-H. Recent progress in near-field nanolithography using light interactions with colloidal particles: From nanospheres to three-dimensional nanostructures. *Nanotechnology* **2019**, *30*, 352002. [CrossRef]
48. Horiuchi, N. Photonic nanojets. *Nat. Phot.* **2012**, *6*, 138–139. [CrossRef]
49. Wannemacher, R. Confocal Laser Scanning Microscopy. In *Encyclopedia of Nanotechnology*; Bhushan, B., Ed.; Springer: Berlin, Germany, 2016.

 *photonics*

Communication

Photonic Nanojet Modulation Achieved by a Spider-Silk-Based Metal–Dielectric Dome Microlens

Ching-Bin Lin [1], Yu-Hsiang Lin [1], Wei-Yu Chen [2] and Cheng-Yang Liu [2,*]

[1] Department of Mechanical and Electro-Mechanical Engineering, Tamkang University, New Taipei City 25137, Taiwan; cblin@mail.tku.edu.tw (C.-B.L.); 607370037@gms.tku.edu.tw (Y.-H.L.)

[2] Department of Biomedical Engineering, National Yang Ming Chiao Tung University, Taipei 112, Taiwan; sam30261014.be09@nycu.edu.tw

* Correspondence: cyliu66@nycu.edu.tw; Tel.: +886-2-28267020

Abstract: The photonic nanojet is a non-resonance focusing phenomenon with high intensity and narrow spot that can serve as a powerful biosensor for in vivo detection of red blood cells, microorganisms, and tumor cells in blood. In this study, we first demonstrated photonic nanojet modulation by utilizing a spider-silk-based metal–dielectric dome microlens. A cellar spider was employed in extracting the silk fiber, which possesses a liquid-collecting ability to form a dielectric dome microlens. The metal casing on the surface of the dielectric dome was coated by using a glancing angle deposition technique. Due to the nature of surface plasmon polaritons, the characteristics of photonic nanojets are strongly modulated by different metal casings. Numerical and experimental results showed that the intensity of the photonic nanojet was increased by a factor of three for the gold-coated dome microlens due to surface plasmon resonance. The spider-silk-based metal-dielectric dome microlens could be used to scan a biological target for large-area imaging with a conventional optical microscope.

Keywords: photonic nanojet; spider silk; dome lens

Citation: Lin, C.-B.; Lin, Y.-H.; Chen, W.-Y.; Liu, C.-Y. Photonic Nanojet Modulation Achieved by a Spider-Silk-Based Metal–Dielectric Dome Microlens. *Photonics* **2021**, *8*, 334. https://doi.org/10.3390/photonics8080334

Received: 21 July 2021
Accepted: 12 August 2021
Published: 14 August 2021

Publisher's Note: MDPI stays neutral with regard to jurisdictional claims in published maps and institutional affiliations.

1. Introduction

Currently, the design of mesoscale photonic devices with high spatial resolution and operation speed opens up prospects for the evolution of novel microscopic and manufacturing technologies [1,2]. According to the principle of optics, the spatial resolution of traditional optical elements is defined by diffraction, and the minimum dimension of the focal spot is more than half the wavelength of the incident light wave. This description signifies that traditional optical elements in the circuit are positioned at an extensive distance from each other compared to the wavelength. Therefore, multitudinous investigations have verified the materialization of a high-intensity optical field restrained in a region when the light wave was focused by a mesoscale dielectric particle [3–14]. This phenomenon is referred to in several academic papers as the photonic nanojet effect. For generating a photonic nanojet, the Mie size parameter ($q = 2\pi r/\lambda$) of a dielectric particle should correspond to $q \sim (2 \ldots 40)\pi$, where r is the particle radius and λ is the operating wavelength [15,16]. In contrast to the case of focusing radiation with a traditional optical lens, a photonic nanojet is formed in the near-scattering area, where the intensity field is a complex spatial structure determined by a superposition of outgoing and decaying evanescent waves. The intensity distribution and localization of a photonic nanojet depends on the shapes and physical properties of the mesoscale dielectric particles and surrounding media. As a rule, the assemblies of mesoscale particles are embedded in a polymer film, which makes it possible to arrange particles in diverse spatial configurations [17–20]. Such a composite polymer film will lead to a multiple increase in the intensity field of a photonic nanojet. However, one of the problems with photonic nanojets in near-field focusing is the short spatial length of the focal region that emerges near the convex surface of classical particle geometry. Hence, the searches for mesoscale particles of other geometric shapes are stimulated to extend the length of the photonic nanojet as far as possible from the particle surface [21,22]. It is also noted that the characteristics (length, width, and peak intensity) of photonic nanojets can be varied by changing the geometric parameters of mesoscale particles such as spheroids, hemispheres, core-shell spheres, toroids, axicons, pyramids, and cuboids [23–31]. However, the fabrication of composite inhomogeneous particles is a problem of great complexity. A natural method is required for assembling mesoscale particular particles in the next step. In natural materials, spider silk possesses several valuable features, including large tensile strength, great toughness, and high elasticity [32–34]. In general, spider silk is enclosed by meshes with nanosized strings and cavities. The appearance of a lyotropic liquid crystalline phase determines the mechanical properties of spider silk. Among spider silks, dragline silk possesses large tensile strength and is used as the skeleton of a web. Natural silk fiber can be utilized for optical guiding, imaging, and sensing applications due to its biocompatibility, bioresorbability, and excellent mechanical properties [35–40]. An optical microlens based on natural silk fiber could therefore be very practical, and needs further demonstration.

In this study, we first theoretically and experimentally demonstrate efficient photonic nanojet modulation of a mesoscale metal–dielectric dome microlens based on natural silk fiber. The silk fibers were directly collected from a daddy long-legs spider for fabricating the dome microlens due to their better homogeneity and mechanical properties. The glancing angle deposition technique was used to coat different metal layers on the dome surface. The inspections of the photonic nanojet modulations were performed by numerical simulations and a laser scanning digital microscope. The critical parameters of the focusing spot for the dome microlens with different metal coatings were studied systematically.

2. Experimental Methods

2.1. Metal–Dielectric Dome Microlens

The cellar spider (*Pholcus phalangioides*), commonly known as the "daddy long-legs", can be found throughout the world, especially in undisturbed low-light locations. People often associate this spider with living in the corners of a rooms near the floor and ceiling. Figure 1a shows a daddy long-legs spider that usually spins its webs large, loose, and flat. A daddy long-legs spider was employed to output a single strand of silk fiber from the major ampullate gland for the experiments. Figure 1b shows the electric reeling system for silk fiber collection under controlled conditions of reeling speed, humidity, and temperature. A silk fiber with a smooth surface, circular cross-section, and uniform material quality was extracted with the reeling process. This silk fiber was a transparent medium, and the refractive index was about 1.55 in visible light region [35]. Due to the high refractive index contrast between silk fiber and the surrounding air, optical multimode guidance can be excited in the silk fiber [37]. After the reeling collection, a silk fiber with a 7 cm length and a 2 μm diameter was fixed at both ends of a specialized holder. A paraffin wax was placed below the bare silk fiber, and the photocurable resin (Everwide FP098) was dropped onto the silk's surface. The refractive indices of the photocurable resin were 1.519, 1.503, and 1.495 for 405 nm, 532 nm, and 671 nm wavelengths, respectively. Due to the obstruction of the paraffin wax, the photocurable resin was concentrated only in one direction and formed a dome shape on the silk fiber. In the initial phase, slight resin drops condensed on the transparent puffs. As resin condensation continued, the puffs enlarged into bumps and finally became periodic dome shapes. The standing time of the photocurable resin determined the dome dimension because the silk fiber had the excellent ability of directional liquid collection. This structural wet-rebuilding ability of silk fiber has been reported in previous scientific literature [33]. Finally, the solidified dielectric dome microlens was obtained by using an ultraviolet oven (OPAS TX-500ST, Ganbow Technology Co., New Taipei City, Taiwan) and curing statically for 12 s. Furthermore, the glancing angle deposition technique was performed to coat different metal nanolayers on the dome surface [41]. Figure 1c shows the sputtering manufacturing process for coating the spider-silk-based metal-dielectric dome microlens. The metal nanolayer could be uniformly deposited on the dome surface because the silk fiber was inclined with respect to the metal target during the sputtering process. A scanning electron microscope (SEM) was employed to obtain the actual images of the dome microlens on the silk fiber. Figure 1d exhibits an SEM image of the spider-silk-based metal-dielectric dome microlens. It was observed that the diameter of the silk fiber was about 2 μm, and the uniformity of dome shape was excellent. In order to verify the metal layer's thickness, the metal–dielectric dome microlens was cleaved by a commercial focused ion beam system (Helios NanoLab 600i, FEI, Hillscoro, OR, USA). It was clarified that the thickness of the metal nano-layer was about 5 nm. The silk-based metal–dielectric dome microlens could be utilized as a plasmonic device.

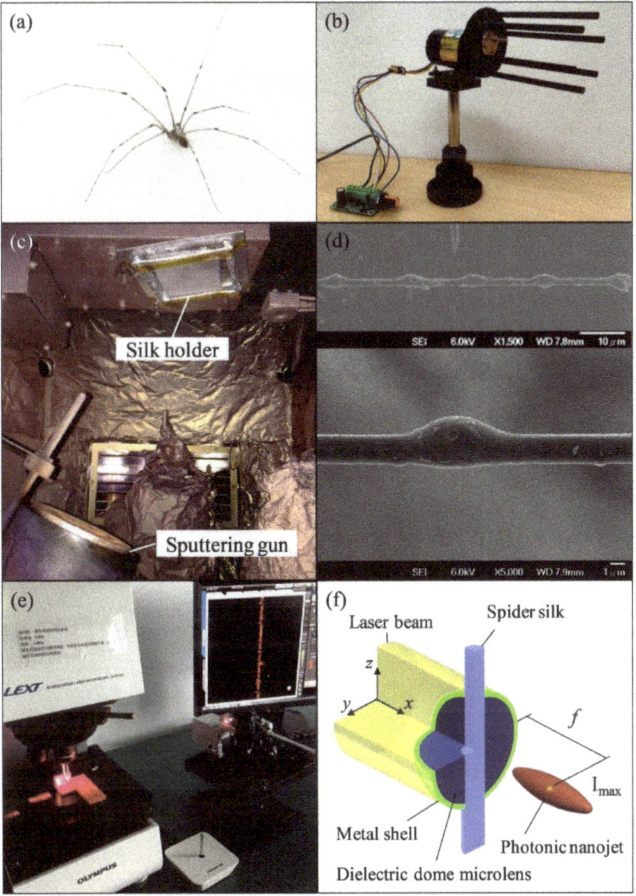

Figure 1. (**a**) A photo of a daddy long-legs spider. (**b**) The electric reeling system for the spider silk. (**c**) The sputtering manufacturing process for coating the spider-silk-based metal–dielectric dome microlens. (**d**) An SEM image of the dome microlens. (**e**) The laser scanning digital microscope system for measuring the dome microlens. (**f**) A schematic diagram of the dome microlens for photonic nanojet modulation.

2.2. Measurement Setup

In the experiments, a commercial laser scanning digital microscope system (LEXT OLS4100, Olympus, Tokyo, Japan) was employed for measuring the optical field intensity of different metal–dielectric dome microlenses [42]. Figure 1e shows the experimental configuration of the measurement system. A diode-pumped solid-state laser with wavelengths of 405 nm, 532 nm, and 671 nm was used for illuminating the dome microlens. The metal-dielectric dome microlens was clamped on a specialized holder and fastened on a three-axis motorized stage for aligning the laser beam. The auto-focus processing, based on Olympus software, was used to obtain accurate cross-section images with a 10 nm height resolution and 120 nm lateral resolution. The experimental images of the field intensity distributions generated by the dome microlens were acquired by using an objective lens (MPLAPON100XLEXT, Olympus, Tokyo, Japan) with a working distance of 0.35 mm, a numerical aperture of 0.95, and a photomultiplier. The technical details of the commercial laser scanning digital microscope can be found on the official website for Olympus. Figure 1f shows the schematic diagram of the metal-dielectric dome microlens for photonic nanojet modulation. The convex side of the metal-dielectric dome microlens was illuminated by the laser beam along the x axis. The photonic nanojet is shown on the right side of the dome microlens. The focal length f is the axial distance from the flat side of the dome microlens to the maximum peak amplitude (I_{max}) along the x axis. The decay length is the axial distance from the I_{max} at which the intensity distribution drops to I_{max}/e along the x axis. The full width at half-maximum (FWHM) is the transverse width between the I_{max} and half-maximum point along the z axis. The finite-difference time-domain (FDTD) method was utilized to build the simulation model of the metal-dielectric dome microlens [43]. The mesh grid in the metal layer region was 1 nm for high accuracy, but the mesh grid in the dielectric and surrounding media was 20 nm for high calculation speed. Perfectly matched boundary layers were implemented along the boundaries of the simulation area. The gold-, silver-, and copper-coating layers had a refractive index of 0.54 + 2.23i, 0.05 + 3.43i, and 1.12 + 2.59i, respectively, at a wavelength of 532 nm [44]. The surrounding medium was air with a refractive index of 1.

3. Results and Discussions

In order to achieve the function of photonic nanojet beam steering, it is important to be able to arbitrarily modulate the focusing property from different metal–dielectric dome microlenses. The numerical and experimental results of the photonic nanojet modulation were verified as indicated below. Figure 2a–d display the numerical results of normalized power flow patterns for the dielectric, gold-coated, silver-coated, and copper-coated dome microlenses. From the amplitude distributions, one can observe that the photonic nanojet was formed close to the surface of the microlens for the dielectric dome microlens, while it moved away from the microlens surface for the metal-coated dome microlenses. In the case of a general sphere, the photonic nanojet only has a short focal length due to the rapid convergence and divergence near the focusing point [11]. The photonic nanojet can be modified by changing the design of the engineered spheres, which leads to a sharp spot size [12]. This concept demonstrated that the dome microlens acted like a ball lens, and the focusing spot was deformed along the propagation direction by the different metal nanolayer coatings. It was also observed that the effect of the metal nanolayers on the photonic nanojet beam shaping was clarified by the power flow patterns. The maximum intensity of the photonic nanojet changed as the metal nanolayers of dome microlenses changed. As seen from the power flow patterns, the intensity of the photonic nanojet was effectively amplified by coating a gold nanolayer on the dome surface, due to the surface plasmon polaritons [45].

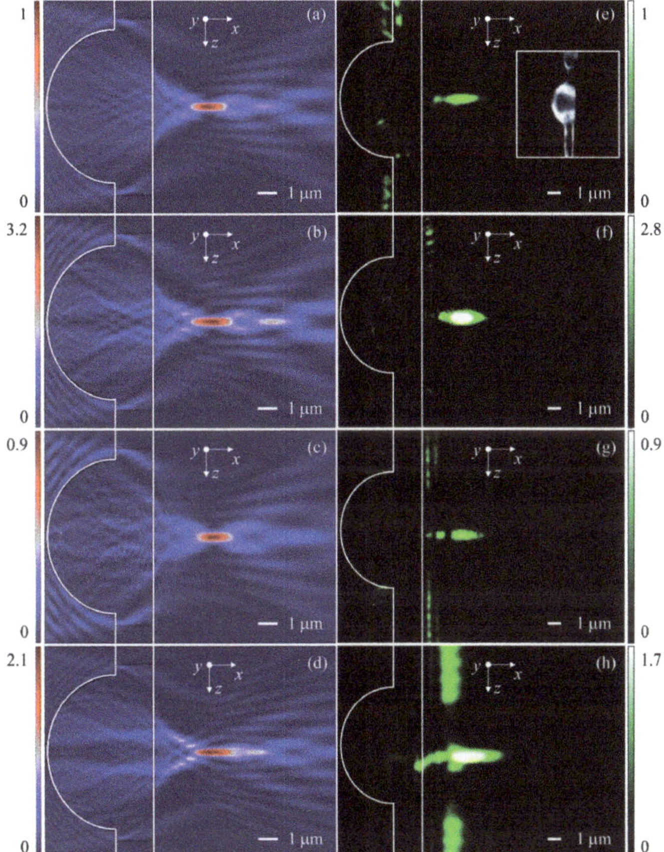

Figure 2. Numerical results of normalized power flow patterns for the (**a**) dielectric, (**b**) gold-coated, (**c**) silver-coated, and (**d**) copper-coated dome microlenses; and raw experimental images of the (**e**) dielectric, (**f**) gold-coated, (**g**) silver-coated, and (**h**) copper-coated dome microlenses. The incident wavelength was 532 nm for all the dome microlenses. The inset is a microphotograph of the spider-silk-based dome microlens.

Figure 2e–h display the direct experimental images of the intensity distribution for different metal-dielectric dome microlenses at a 532 nm wavelength for verification. Compared with the simulation results, the focusing effect of the photonic nanojet is exhibited clearly in the raw images. Apparently, the field intensity distributions of the experiments were largely in agreement with the simulation results. As shown in Figure 2a,e, the dielectric dome microlens generated the photonic nanojet, the shape of which behaved in a manner similar to a stiletto knife. The field intensity distributions around the focusing point were nearly parallel, and therefore formed a narrow strip. It can be seen in Figure 2f that the focusing intensity of the gold-coated dome microlens was enhanced significantly due to the surface plasmon resonance. The photonic nanojet's focus with the surface plasmon resonance was almost three times the field intensity of the dielectric dome microlens. Controllable photonic nanojet formation excited by plasmonic effects can be described by the dispersion relation of the surface plasmon polaritons [45,46]. The experimental image in Figure 2g demonstrates that the surface plasmon absorption excited on the silver layer caused the intensity reduction of the photonic nanojet for the silver-coated dome microlens. In Figure 2h, the dispersion effect on the surface of the copper-coated dome microlens can be observed due to the surface plasmon scattering. The surface-dependent reflectance arose from the copper layer, and the optical beam was dispersed from the layer surface.

To quantitatively estimate the quality of the photonic nanojet, we determined several critical parameters from the field intensity distribution. Figure 3 shows the critical parameters as a function of the incident wavelength for the dome microlenses with different metal coatings. The focal lengths at a 532 nm wavelength were measured to be 2.27 μm, 2.66 μm, 2.75 μm, and 2.88 μm for the dielectric, gold-coated, silver-coated, and copper-coated dome microlenses, respectively. The focal length increased as the incident wavelength increased. Compared to the dielectric dome microlens, the focal length was increased by 27% for the copper-coated dome microlens. The focal length could be adjusted by varying the metal layer. When the dome microlens was coated with a metal layer, the focal length was large enough to meet the working distance for microlens-aided imaging [2]. In Figure 3b, the decay lengths at a 532 nm wavelength were measured to be 1.93 μm, 2.29 μm, 2.16 μm, and 2.23 μm for the dielectric, gold-coated, silver-coated, and copper-coated dome microlenses, respectively. It was observed that the decay length had a maximum value at an incident wavelength of 532 nm for all the metal-dielectric dome microlenses. In Figure 3c, the transverse FWHMs at a 532 nm wavelength were measured to be 192 nm, 215 nm, 232 nm, and 263 nm for the dielectric, gold-coated, silver-coated, and copper-coated dome microlenses, respectively. The transverse FWHM of the photonic nanojet decreased as the incident wavelength decreased. As the incident wavelength decreased to 405 nm, the corresponding FWHM became significantly less than 0.5λ via the experimental verifications. The suggested metal-dielectric dome microlenses showed great potential for far-field super-resolution lithography and imaging applications [42].

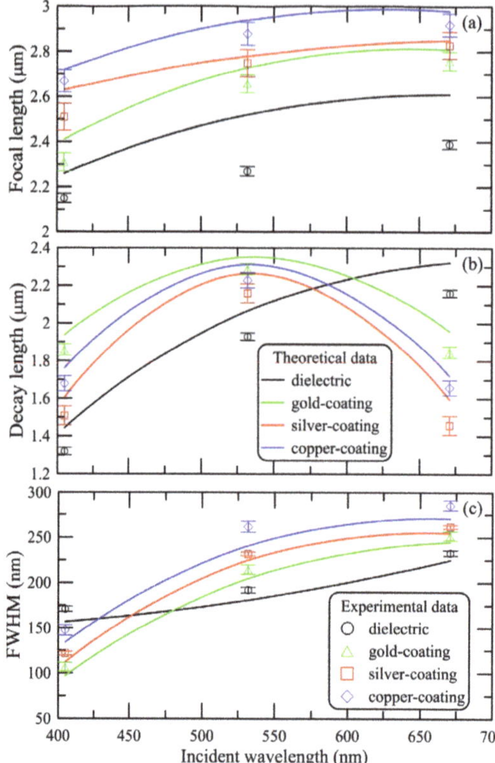

Figure 3. Critical parameters as a function of incident wavelength for the dome microlenses with different metal coatings: (**a**) focal length, (**b**) decay length, and (**c**) FWHM.

Figure 4a shows the normalized intensity distributions of photonic nanojets for the different metal-dielectric dome microlenses along the propagation axis (*x* axis). The origin of the longitudinal profile along the propagation axis was located at the flat surface of the dome microlenses. All intensity profiles were normalized to the intensity profile for the dielectric dome microlens. The photonic nanojet generated by dome microlenses emerged in the form of a Gaussian distribution with an exponentially decaying trail. The length of the photonic nanojet increased as the dome microlens was coated with metal layer, and the field intensity was enhanced as well. The maximum intensity increased by 180% for the gold coating, although the maximum intensity decreased by 10% for the silver coating. Depending on the metal layer, we observed that not only the intensity of the photonic nanojet was enhanced, but the effective length of the photonic nanojet also was elongated. The engineered metal–dielectric dome microlens is expected to provide high concentration and low divergence in the focusing point [24].

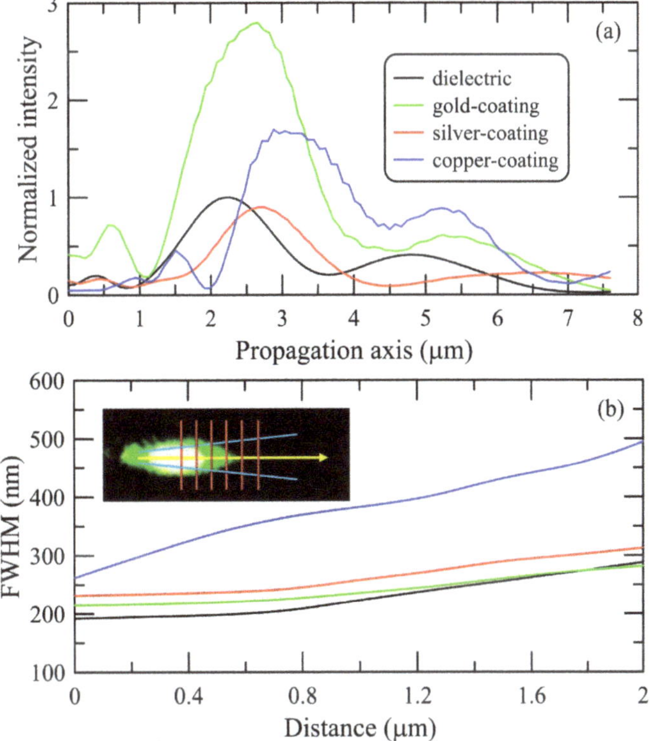

Figure 4. (**a**) Normalized intensity distributions of photonic nanojets for different metal-dielectric dome microlenses along the propagation axis (*x* axis). (**b**) FWHM as a function of the propagation distance for photonic nanojets for the different metal-dielectric dome microlenses. The inset indicates that the yellow arrow is the propagation direction, the red lines are the positions of serial cross-sections for the FWHM, and the blue lines are the straight fitting lines.

Figure 4b shows the FWHM as a function of the propagation distance for the photonic nanojets with different metal–dielectric dome microlenses. The origin of the propagation distance corresponded to the point of maximal peak amplitude. The slope of the straight fitting line was used to determine the divergence angle of the photonic nanojets [7]. The divergence angles at a 532 nm wavelength were measured to be 6.3°, 2.6°, 3.8°, and 7.2° for the dielectric, gold-coated, silver-coated, and copper-coated dome microlenses, respectively. The proposed metal-coated dome microlens can work with a low divergence angle and long focal length, which is not possible for dielectric spherical microlenses. It was apparent that the divergence angle of the designed dome microlens was sensitive to the refractive index of the surrounding medium. When the dome microlens was coated with a metal layer, the refractive index of the surrounding medium could be found indirectly by measuring the divergence angle. Both the above-mentioned experimental and simulation results confirmed that the metal-dielectric dome microlens with flexible photonic nanojet modulation was suitable for a plasmonic sensor of the refractive index [47].

4. Conclusions

In this work, photonic nanojet modulation based on a metal-dielectric dome microlens was first theoretically and experimentally demonstrated with an extended focal length, narrow beam waist, long effective length, and low divergence angle. The directional liquid collection capability of wet-rebuilt silk fiber was utilized for the formation of the dielectric dome microlens. The dielectric dome microlens was coated with different metal layers by using the glancing angle deposition technique. Through FDTD simulation and experimental analysis, we concluded that the improvement of the photonic nanojet was attributed to the nature of surface plasmon polaritons, which caused a high concentration near the metal-dielectric interface. The gold-coated dome microlens had an intensity enhancement of about three times due to surface plasmon resonance. The focal length of the photonic nanojet was increased by 27% for the copper-coated dome microlens. A minimum divergence angle of $2.6°$ was achieved by the gold-coated dome microlens. Moreover, the proposed metal-dielectric dome microlens showed compatibility with the adjacent wavelengths and a spot width less than half-wavelength. These kinds of plasmonic microlenses have great potential in far-field flexible parallel lithography with a sub-wavelength line width.

Author Contributions: Conceptualization, C.-B.L. and C.-Y.L.; methodology, C.-B.L. and C.-Y.L.; software, C.-Y.L. and W.-Y.C.; validation, C.-Y.L. and W.-Y.C.; formal analysis, C.-B.L. and Y.-H.L.; resources, C.-Y.L.; data curation, Y.-H.L. and W.-Y.C.; writing—original draft preparation, C.-Y.L.; writing—review and editing, C.-B.L. and C.-Y.L.; visualization, C.-Y.L. and W.-Y.C.; supervision, C.-B.L. and C.-Y.L.; project administration, C.-Y.L. and W.-Y.C.; funding acquisition, C.-Y.L. All authors have read and agreed to the published version of the manuscript.

Funding: This research was funded by the Ministry of Science and Technology of Taiwan (MOST 108-2221-E-010-012-MY3, MOST 109-2923-E-010-001-MY2) and the Yen Tjing Ling Medical Foundation (CI-110-28).

Institutional Review Board Statement: Not applicable.

Informed Consent Statement: Not applicable.

Data Availability Statement: The data that support the findings of this study are available from the corresponding author upon reasonable request.

Conflicts of Interest: The authors declare no conflict of interest.

References

1. Allen, K.; Farahi, N.; Li, Y.; Limberopoulos, N.; Walker, D., Jr.; Urbas, A.; Liberman, V.; Astratov, V. Super-resolution mi-croscopy by movable thin-films with embedded microspheres: Resolution analysis. *Ann. Phys.* **2015**, *527*, 513–522. [CrossRef]
2. Chen, L.; Zhou, Y.; Li, Y.; Hong, M. Microsphere enhanced optical imaging and patterning: From physics to applications. *Appl. Phys. Rev.* **2019**, *6*, 021304. [CrossRef]
3. Chen, Z.; Taflove, A.; Backman, V. Photonic nanojet enhancement of backscattering of light by nanoparticles: A potential novel visible-light ultramicroscopy technique. *Opt. Express* **2004**, *12*, 1214–1220. [CrossRef] [PubMed]
4. Li, X.; Chen, Z.; Taflove, A.; Backman, V. Optical analysis of nanoparticles via enhanced backscattering facilitated by 3-D photonic nanojets. *Opt. Express* **2005**, *13*, 526–533. [CrossRef] [PubMed]
5. Ferrand, P.; Wenger, J.; Devilez, A.; Pianta, M.; Stout, B.; Bonod, N.; Popov, E.; Rigneault, H. Direct imaging of photonic nanojets. *Opt. Express* **2008**, *16*, 6930–6940. [CrossRef] [PubMed]
6. Kim, M.; Scharf, T.; Mühlig, S.; Rockstuhl, C.; Herzig, H.P. Engineering photonic nanojets. *Opt. Express* **2011**, *19*, 10206–10220. [CrossRef]

7. McCloskey, D.; Wang, J.; Donegan, J. Low divergence photonic nanojets from Si3N4 microdisks. *Opt. Express* **2012**, *20*, 128–140. [CrossRef]
8. Han, L.; Han, Y.; Wang, J.; Cui, Z. Internal and near-surface electromagnetic fields for a dielectric spheroid illuminated by a zero-order Bessel beam. *J. Opt. Soc. Am. A* **2014**, *31*, 1946–1955. [CrossRef]
9. Liu, C.; Chen, C. Characterization of photonic nanojets in dielectric microdisks. *Phys. E Low Dimens. Syst. Nanostruct.* **2015**, *73*, 226–234. [CrossRef]
10. Liu, C.; Lin, F. Geometric effect on photonic nanojet generated by dielectric microcylinders with non-cylindrical cross-sections. *Opt. Commun.* **2016**, *380*, 287–296. [CrossRef]
11. Wu, M.; Chen, R.; Ling, J.; Chen, Z.; Chen, X.; Ji, R.; Hong, M. Creation of a longitudinally polarized photonic nanojet via an engineered microsphere. *Opt. Lett.* **2017**, *42*, 1444–1447. [CrossRef] [PubMed]
12. Zhou, Y.; Gao, H.; Teng, J.; Luo, X.; Hong, M. Orbital angular momentum generation via a spiral phase microsphere. *Opt. Lett.* **2018**, *43*, 34–37. [CrossRef]
13. Liu, C.; Yeh, M. Experimental verification of twin photonic nanojets from a dielectric microcylinder. *Opt. Lett.* **2019**, *44*, 3262–3265. [CrossRef] [PubMed]
14. Minin, I.; Liu, C.; Geints, Y.; Minin, O. Recent advantages in integrated photonic jet-based photonics. *Photonics* **2020**, *7*, 41. [CrossRef]
15. Heifetz, A.; Kong, S.; Sahakian, A.; Taflove, A.; Backman, V. Photonic nanojets. *J. Comput. Theor. Nanosci.* **2009**, *6*, 1979–1992. [CrossRef]
16. Luk'Yanchuk, B.; Paniagua-Dominguez, R.; Minin, I.V.; Minin, O.V.; Wang, Z. Refractive index less than two: Photonic nanojets yesterday, today and tomorrow. *Opt. Mater. Express* **2017**, *7*, 1820–1847. [CrossRef]
17. Wang, Z.; Guo, W.; Luk'yanchuk, B.; Whitehead, D.; Li, L.; Liu, Z. Optical near-field interaction between neighbouring micro/nano-particles. *J. Laser Micro Nanoeng.* **2008**, *3*, 14–18. [CrossRef]
18. Rizzato, S.; Primiceri, E.; Monteduro, A.; Colombelli, A.; Leo, A.; Manera, M.; Rella, R.; Maruccio, G. Interaction-tailored or-ganization of large-area colloidal assembliesr. *Beilstein J. Nanotechnol.* **2018**, *9*, 1582–1593. [CrossRef]
19. Geints, Y.; Panina, E.; Zemlyanov, A. Collective effects in the formation of an ensemble of photonic nanojets by an ordered microassembly of dielectric microparticles. *Atmos. Ocean. Opt.* **2019**, *32*, 113–119. [CrossRef]
20. Geints, Y.; Zemlyanov, A.; Panina, E. Peculiarities of the formation of an ensemble of photonic nanojets by a micro-assembly of conical particles. *Quantum Electron.* **2019**, *49*, 210–215. [CrossRef]
21. Geints, Y.; Zemlyanov, A.; Panina, E. Photonic nanojet calculations in layered radially inhomogeneous micrometer-sized spherical particles. *J. Opt. Soc. Am. B* **2011**, *28*, 1825–1830. [CrossRef]
22. Shen, Y.; Wang, L.; Shen, J. Ultralong photonic nanojet formed by a two-layer dielectric microsphere. *Opt. Lett.* **2014**, *39*, 4120–4123. [CrossRef] [PubMed]
23. Pacheco-Peña, V.; Beruete, M.; Minin, I. Terajets produced by dielectric cuboids. *Appl. Phys. Lett.* **2014**, *105*, 084102. [CrossRef]
24. Liu, C.; Hsiao, K. Direct imaging of optimal photonic nanojets from core-shell microcylinders. *Opt. Lett.* **2015**, *40*, 5303–5306. [CrossRef]
25. Geints, Y.; Zemlyanov, A.; Panina, E. Microaxicon-generated photonic nanojets. *J. Opt. Soc. Am. B* **2015**, *32*, 1570–1574. [CrossRef]
26. Minin, I.V.; Minin, O.V.; Geints, Y. Localized EM and photonic jets from non-spherical and non-symmetrical dielectric mesoscale objects: Brief review. *Ann. Phys.* **2015**, *527*, 491–497. [CrossRef]
27. Degtyarev, S.; Porfirev, A.; Khonina, S. Photonic nanohelix generated by a binary spiral axicon. *Appl. Opt.* **2016**, *55*, B44–B48. [CrossRef] [PubMed]
28. Geints, Y.; Minin, I.V.; Panina, E.; Zemlyanov, A.; Minin, O.V. Comparison of photonic nanojets key parameters produced by nonspherical microparticles. *Opt. Quantum Electron.* **2017**, *49*, 118. [CrossRef]
29. Mahariq, I.; Giden, I.; Kurt, H.; Minin, O.V.; Minin, I.V. Strong electromagnetic field localization near the surface of hemicy-lindrical particles. *Opt. Quant. Electron.* **2018**, *50*, 423. [CrossRef]
30. Zhang, B.; Hao, J.; Shen, Z.; Wu, H.; Zhu, K.; Xu, J.; Ding, J. Ultralong photonic nanojet formed by dielectric microtoroid structure. *Appl. Opt.* **2018**, *57*, 8331–8337. [CrossRef]
31. Liu, X.; Zhou, H.; Yang, M.; Xie, Z.; Han, Q.; Gou, J.; Wang, J. Photonic nanojets with ultralong working distance and narrowed beam waist by immersed engineered dielectric hemisphere. *Opt. Express* **2020**, *28*, 33959–33970. [CrossRef]
32. Omenetto, F.; Kaplan, D. A new route for silk. *Nat. Photonics* **2008**, *2*, 641–643. [CrossRef]
33. Zheng, Y.; Bai, H.; Huang, Z.; Tian, X.; Nie, F.; Zhao, Y.; Zhai, J.; Jiang, L. Directional water collection on wetted spider silk. *Nat. Cell Biol.* **2010**, *463*, 640–643. [CrossRef]
34. Rising, A.; Johansson, J. Toward spinning artificial spider silk. *Nat. Chem. Biol.* **2015**, *11*, 309–315. [CrossRef]
35. Little, D.; Kane, D. Image contrast immersion method for measuring refractive index applied to spider silks. *Opt. Express* **2011**, *19*, 19182–19189. [CrossRef] [PubMed]
36. Huby, N.; Vié, V.; Renault, A.; Beaufils, S.; Lefèvre, T.; Paquet-Mercier, F.; Pézolet, M.; Bêche, B. Native spider silk as a biological optical fiber. *Appl. Phys. Lett.* **2013**, *102*, 123702. [CrossRef]
37. Applegate, M.; Perotto, G.; Kaplan, D.; Omenetto, F. Biocompatible silk step-index optical waveguides. *Biomed. Opt. Express* **2015**, *6*, 4221–4227. [CrossRef] [PubMed]
38. Kujala, S.; Mannila, A.; Karvonen, L.; Kieu, K.; Sun, Z. Natural Silk as a Photonics Component: A Study on Its Light Guiding and Nonlinear Optical Properties. *Sci. Rep.* **2016**, *6*, 22358. [CrossRef] [PubMed]
39. Monks, J.; Yan, B.; Hawkins, N.; Vollrath, F.; Wang, Z. Spider Silk: Mother Nature's Bio-Superlens. *Nano Lett.* **2016**, *16*, 5842–5845. [CrossRef] [PubMed]

40. Tow, K.; Chow, D.; Vollrath, F.; Dicaire, I.; Gheysens, T.; Thevenaz, L. Exploring the Use of Native Spider Silk as an Optical Fiber for Chemical Sensing. *J. Light. Technol.* **2017**, *36*, 1138–1144. [CrossRef]
41. Wakefield, N.; Sorge, J.; Taschuk, M.; Bezuidenhout, L.; Brett, M.; Sit, J. Control of the principal refractive indices in biaxial metal oxide films. *J. Opt. Soc. Am. A* **2011**, *28*, 1830–1840. [CrossRef]
42. Zhu, H.; Yan, B.; Zhou, S.; Wang, Z.; Wu, L. Synthesis and super-resolution imaging performance of a refrac-tive-index-controllable microsphere superlens. *J. Mater. Chem. C* **2015**, *3*, 10907–10915. [CrossRef]
43. Taflove, A.; Hagness, S. *Computational Electrodynamics: The Finite Difference Time Domain Method*; Artech House: Norwood, MA, USA, 2005.
44. Johnson, P.; Christy, R. Optical Constants of the Noble Metals. *Phys. Rev. B* **1972**, *6*, 4370–4379. [CrossRef]
45. Ju, D.; Pei, H.; Jiang, Y.; Sun, X. Controllable and enhanced nanojet effects excited by surface plasmon polariton. *Appl. Phys. Lett.* **2013**, *102*, 171109. [CrossRef]
46. Minin, I.V.; Minin, O.V.; Glinskiy, I.; Khabibullin, R.; Malureanu, R.; Lavrinenko, A.; Yakubovsky, D.; Arsenin, A.; Volkov, V.; Ponomarev, D. Plasmonic nanojet: An experimental demonstration. *Opt. Lett.* **2020**, *45*, 3244–3247. [CrossRef]
47. Liu, X.; Wang, J.; Gou, J.; Ji, C.; Cui, G. Optical Properties and Sensing Performance of Au/SiO$_2$ Triangles Arrays on Reflection Au Layer. *Nanoscale Res. Lett.* **2018**, *13*, 335. [CrossRef]

MDPI

Article

Photonic Jet-Shaped Optical Fiber Tips versus Lensed Fibers

Djamila Bouaziz [1,2,3,*], Grégoire Chabrol [1,4], Assia Guessoum [2], Nacer-Eddine Demagh [2] and Sylvain Lecler [1,3]

[1] ICube Research Institute, University of Strasbourg, CNRS, 67412 Illkirch, France;
g.chabrol@ecam-strasbourg.eu (G.C.); sylvain.lecler@insa-strasbourg.fr (S.L.)

[2] Applied Optics Laboratory, Ferhat Abbas University Setif 1, Setif 19000, Algeria;
assiademagh@univ-setif.dz (A.G.); ndemagh@univ-setif.dz (N.-E.D.)

[3] INSA de Strasbourg, 67000 Strasbourg, France

[4] ECAM Strasbourg-Europe, 67300 Schiltigheim, France

* Correspondence: djamila.bouaziz@etu.unistra.fr

Abstract: Shaped optical fiber tips have recently attracted a lot of interest for photonic jet light focusing due to their easy manipulation to scan a sample. However, lensed optical fibers are not new. This study analyzes how fiber tip parameters can be used to control focusing properties. Our study shows that the configurations to generate a photonic jet (PJ) can clearly be distinguished from more classical-lensed fibers focusing. PJ is a highly concentrated, propagative light beam, with a full width at half maximum (FWHM) that can be lower than the diffraction limit. According to the simulations, the PJs are obtained when light is coupled in the guide fundamental mode and when the base diameter of the microlens is close to the core diameter. For single mode fibers or fibers with a low number of modes, long tips with a relatively sharp shape achieve PJ with smaller widths. On the contrary, when the base diameter of the microlens is larger than the fiber core, the focus point tends to move away from the external surface of the fiber and has a larger width. In other words, the optical system (fiber/microlens) behaves in this case like a classical-lensed fiber with a larger focus spot size. The results of this study can be used as guidelines for the tailored fabrication of shaped optical fiber tips according to the targeted application.

Keywords: photonic jet; nanojet; waveguide; microlens; fiber tip

Citation: Bouaziz, D.; Chabrol, G.; Guessoum, A.; Demagh, N.-E.; Lecler, S. Photonic-Jet Shaped Optical Fiber Tips versus Lensed Fibers. *Photonics* **2021**, *8*, 373. https://doi.org/10.3390/photonics8090373

Received: 6 August 2021
Accepted: 29 August 2021
Published: 7 September 2021

Publisher's Note: MDPI stays neutral with regard to jurisdictional claims in published maps and institutional affiliations.

1. Introduction

The term photonic jet (PJ) was first coined by Z. Chen et al. in 2004 when Modelling cylindrical dielectric structures under plane wave illumination; simulations showed that light can be concentrated at the mesoscale with a full width at half maximum (FWHM) lower than the diffraction limit ($<\lambda/2$, where λ is the free space wavelength), and with a power density significantly higher than the incident wave [1–3]. The key properties of the PJ include: its position (working distance W_D: distance from the tip end to the maximum intensity of the photonic jet), FWHM, light intensity peak value and intensity decay length. The determination of these properties has been the aim of numerous theoretical and experimental studies [4–9]. The properties of the PJ depend on the material refractive indexes, the dielectric object geometric shape and size, and on the incidence wavelength [2,10,11]. PJ can be obtained using cyilnders, spheres or non-spherical dielectric objects [11]. The interest of PJ has been demonstrated for applications such as enhanced Raman scattering [12–14], fluorescence enhancement [15,16], improvement in optical-disk data storage technology [17,18], in sub-micrometer laser etching process on semiconductors and metals [19–22], to achieve optical tweezers [23] and many other applications [24].

Unfortunately, the technique is not easy to implement in some industrial processes such as lithography or spectral analysis, where the PJ must scan a sample with a complex pattern. For such applications, the sphere-like object cannot be directly placed on the sample. A mechanical micro-holder or optical tweezers are required to manipulate it accurately [19]. For these reasons, more recently, attention has been paid to PJ generation with waveguides [25] and optical fibers; solutions have been proposed to put the sphere at the end of a hollow-core optical fiber [26,27] or to use an optical fiber with a shaped tip [21,28–31]. Shaped fiber tips have several advantages: easy to move, no necessary contact with the sample and possibility to collect the backscattered light. Owing to the aforementioned arguments, this technique can become a major solution in industrial processes and characterization such as sub-micron laser processing or high-resolution spectral analysis. Nevertheless, the concept of shaped fiber tips has already been known as microlensed fibers for the past 40 years. Initially, microlensed fibers were extensively exploited to improve the coupling between light sources and optical fibers [32–36]. Microlensed fibers are also used in Optical Coherence Tomography OCT [37], scanning microscopy and spectroscopy [38]. Lensed fibers have been fabricated using a large range of methods including: chemical etching [39], laser heating [40] and electric arc discharge melting [41].

The aim of the present work is to investigate shaped optical fiber tips to identify what distinguishes photonic jet generation from classical-lensed fibers focusing. This study also makes it possible to control the properties of the PJ by optimizing the different parameters of a shaped optical fiber tip.

2. Numerical Approach

First a multimode silica optical fiber with a low number of modes is considered for 2D simulation: core diameter $2a = 20$ µm and cladding diameter of 125 µm (20/125 µm fiber), cladding and core refractive index, 1.44 and 1.457 respectively, corresponding to a numerical aperture (NA) around 0.22. The fiber is exited at a wavelength of 1064 nm by its fundamental mode, approximated by a Gaussian function. The propagation direction is set along the x-axis. The source wave is linearly polarized along the z-axis (see axes in Figure 1). To describe the tip shape, the standard form of the rational quadratic Bézier curve is used, set by the tip height (h), the base diameter (D) and the Bézier weight (W_0) [31]. The tip is considered to have the same refractive index as the fiber core. This minimizes the reflection. Light propagate along the fiber, then through the shaped fiber tip and then in the air.

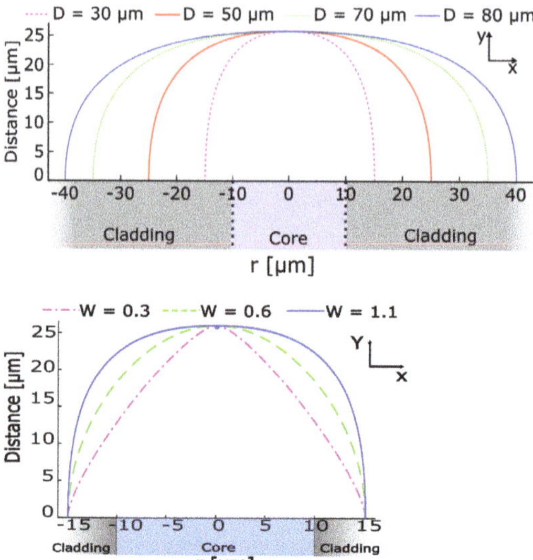

Figure 1. Tip shapes representation. Height $h = 26$ µm. (**Top**) $W_0 = 1.1$, and 4 base diameters D: 30 µm, 50 µm, 70 µm, 80 µm and (**down**) $D = 30$ µm and 3 Bézier weights W_0: 0.3, 0.6, 1.1. Core diameter $2a = 20$ µm.

The electric field distribution is obtained numerically with a 2D finite element method (FEM), COMSOL Multiphysics©, using a free triangular mesh having a maximum mesh size of $\lambda/10$. Scattering boundary conditions are used along the fiber cladding and perfectly matched layers (PMLs) around the free space surrounding the tip.

As illustrated in Figure 1-top, several microlens base diameters (30 µm, 50 µm, 70 µm and 80 µm) were considered in order to study their influence on the PJ using the same shape Bézier weight $W_0 = 1.1$ and height h = 26 µm. The influence of the tip height (h) and sharpness (described by the Bézier weight W_0, see Figure 1-down) have also been studied.

3. Results

Figure 2a depicts the simulation of the intensity ($|E|^2$) distribution inside and outside the shaped optical fiber tip with a base diameter $D = 30$ µm, height $h = 26$ µm, weight $W_0 = 1.1$. A PJ is produced with a full width at half maximum (FWHM) of 1.2 µm and at a working distance W_D of 24 µm.

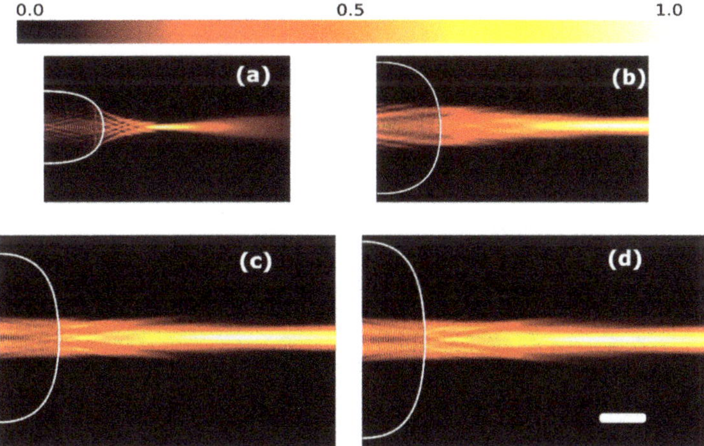

Figure 2. Photonic jet generation (intensity $|E|^2$) for a 20/125 silica optical fiber, $n_g = 1.440$ and $n_c = 1.457$ (NA of 0.22). Excitation by the fundamental mode at a $\lambda = 1064$ nm. Tips with $h = 26$ µm, $W_0 = 1.1$, (**a**) $D = 30$ µm, (**b**) $D = 50$ µm, (**c**) $D = 70$ µm, (**d**) $D = 80$ µm. White scale bars represent 20 µm.

More generally, Figure 2 shows the evolution of the 2D simulated intensity distribution when the tip base diameters increases from 30 µm up to 80 µm. For each configuration the maximum electrical field value outside the tip (for an unitary incident fundamental mode), the corresponding FWHM (evaluate on the intensity) and working distance are given in Table 1.

When the tip diameter increases, the PJ position (W_D) is further, its electrical field maximum decreases and FWHM increases. Figure 2d corresponds to the longer and the largest PJ, with a working distance 200% longer than in Figure 2a. The FWHM is 260% larger and the electric field norm is 46% lower than the PJ obtained with the lowest base diameter.

Table 1. Focus parameters for several fibers and lens diameters D, $\lambda = 1064\,\text{nm}$. Unitary fundamental mode excitation.

Microlens Base Diameter [µm]	FWHM [µm]	W_D [µm]	Electrical Field [V/m]
Fiber 20/125, $NA = 0.22$, $h = 26\,\text{µm}$, $W_0 = 1.1$			
30	1.2	24.6	3.08
50	3.0	65.6	2.08
70	3.9	73.6	1.73
80	4.2	74.7	1.66
Fiber 50/125, $NA = 0.22$, $h = 36\,\text{µm}$, $W_0 = 0.7$			
60	1.1	41	4.63
70	1.2	57.1	4.51
80	1.3	76	4.45
110	2.5	153	3.40
Fiber 10/125, $NA = 0.12$, $h = 14\,\text{µm}$, $W_0 = 1.1$			
15	1.0	9.4	2.76
17	1.1	11.9	2.60
19	1.2	17.7	2.46
21	1.7	23.2	2.12

The corresponding electric field norm $|E|$ profile on the photonic jet planes has been plotted for the 20 µm core fiber but also for optical fibers with core diameter 50 µm (NA = 0.22) and 10 µm (NA = 0.12). Compared with the 20/125 fiber, a longer and not so sharp tip ($h = 36\,\text{µm}$, $W_0 = 0.7$) is required for the 50 µm core fiber and inversely, a smaller and sharper tip ($h = 14\,\text{µm}$, $W_0 = 1.1$) is required for the 10 µm core fiber, thus justifying our choices. For the three fibers, the simulated base diameters have been chosen to be first close to the core diameter and then larger. Figure 3 illustrates the corresponding electric field norm $|E|$ profile on the photonic jet planes. For the three fibers, a PJ is obtained when the microlens base diameter is close to the core diameter. The trend of the simulations are in good agreement: when the base diameter of the tip increases, the maximum electric field decreases, the PJ FWHM and working distance increase. In other words, the optical system (fiber/microlens) acts more and more as a classical-lensed fiber.

We now consider the influence of microlens Bézier weight (sharpness) and height on the focusing properties. As illustrated in Figure 4a with the 20/125 silica fiber, an increase of the Bézier weight (blunter tip) pushes the PJ (W_D) away from the fiber tip end and increases the PJ FWHM (as a lens having a larger focal distance would). The narrower PJ FWHM, around 0.55 µm, is obtained for the the lower weight $W_0 = 0.3$ (sharper tip) at a working distance W_D of 3 µm. Compared with $W_0 = 1.1$, the PJ width is divided by 2. Thus, the present investigation confirms that the Bézier curvature has a primordial role in controlling the focusing parameters.

In Figure 4c,d, we can also observe that the FWHM and W_D decrease with the increase of the tip height.

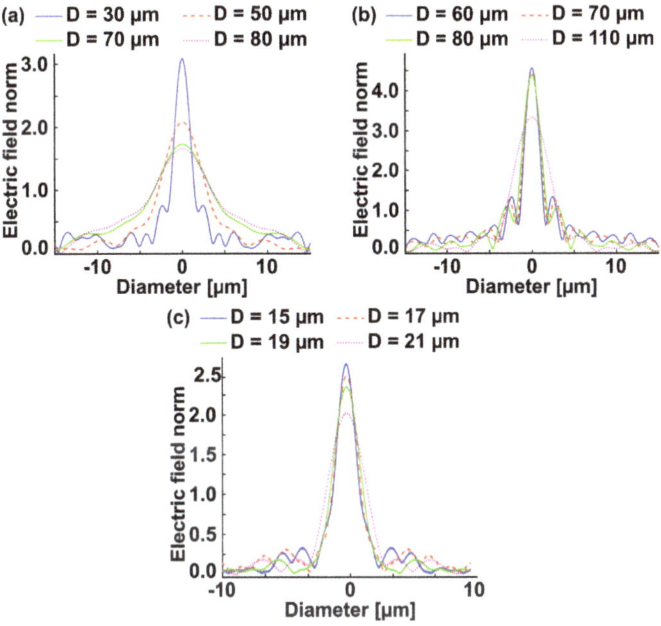

Figure 3. Electric field |E| profile on the PJ planes. The different colors represent the different base diameter D of the microlenses for (**a**) a 20/125 silica fiber ($h = 26$ μm, $W_0 = 1.10$, $NA = 0.22$). (**b**) a 50/125 silica fiber ($h = 36$ μm, $W_0 = 0.7$, $NA = 0.22$). (**c**) a 10/125 silica fiber ($h = 14$ μm, $W_0 = 1.1$, $NA = 0.12$). Fundamental mode excitation at $\lambda = 1064$ nm.

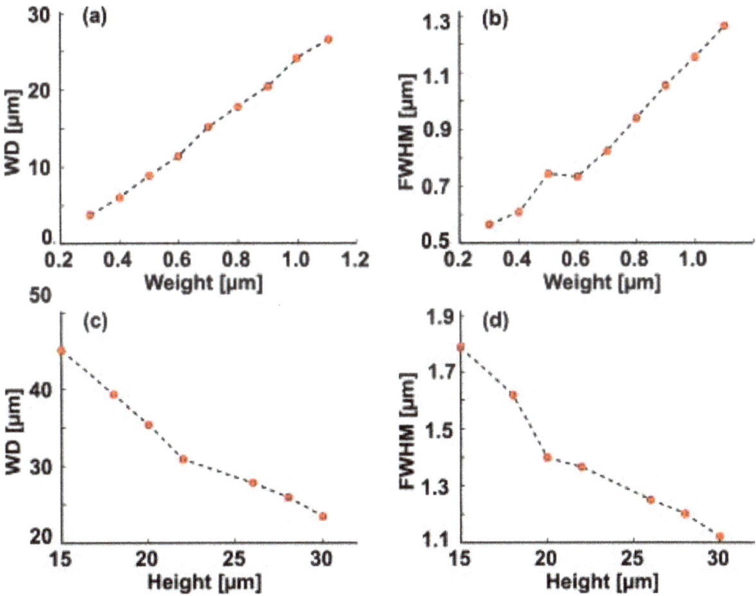

Figure 4. Working distance W_D (**a,c**) and FWHM (**b,d**) variations as a function of Bézier weight (**a,b**) and height (**c,d**) of the shaped tip for the 20/125 fiber ($NA = 0.22$). Default value: $D = 32$ μm, $h = 26$ μm and $W_0 = 1.1$.

4. Conclusions

The present study was designed to control photonic jet generation by assessing the effect of the optical fiber tip parameters and to clarify the differences between classical fiber lens and the shaped fiber tips making possible to obtain PJ. A first already known difference is that PJ is only due to the fundamental mode, namely, the other modes being off-axis focused [31]. We show in this paper that a second important difference is that PJs occur when the microlens base diameter is close to the fiber core diameter, whereas classical fiber lenses stand generally on the whole optical fiber cladding. When the tip is shaped by thermoforming, this can explain why a PJ is easier to obtain with multimode fibers (having a core diameter close to the optical cladding one). PJs achieved using a shaped fiber tip may be used to scan a sample with a complex pattern and also collect the backscattered light with the same fiber. They are, therefore, a promising solution for many applications from laser nano-abalation to local spectrocopy.

Author Contributions: Main investigations, D.B.; formal analysis, A.G. and G.C.; supervision, S.L. and N.-E.D. All authors have read and agreed to the published version of the manuscript.

Funding: This research was funded by SATT Conectus; PHC TASSILI Cooperation Program 19MDU217.

Institutional Review Board Statement: Not applicable, the study did not involve humans or animals.

Informed Consent Statement: Not applicable. Studies not involving humans.

Data Availability Statement: Data supporting reported results are stored in ICube Lab.

Acknowledgments: The authors thank Amandine Elchinger for her useful comments in the editing of the manuscript.

Conflicts of Interest: The authors declare no conflict of interest.

References

1. Chen, Z.; Taflove, A.; Backman, V. Photonic Nanojet Enhancement of Backscattering of Light by Nanoparticles: A Potential Novel Visible-Light Ultramicroscopy Technique. *Opt. Express* **2004**, *12*, 1214–1220. [CrossRef] [PubMed]
2. Lecler, S.; Takakura, Y.; Meyrueis, P. Properties of a Three-Dimensional Photonic Jet. *Opt. Lett.* **2005**, *30*, 2641–2643. [CrossRef] [PubMed]
3. Heifetz, A.; Kong, S.C.; Sahakian, A.V.; Taflove, A.; Backman, V. Photonic Nanojets. *Rev. Mat. Iberoam.* **2009**, *6*, 1979–1992. [CrossRef] [PubMed]
4. Grojo, D.; Sandeau, N.; Boarino, L.; Constantinescu, C.; De Leo, N.; Laus, M.; Sparnacci, K. Bessel-like Photonic Nanojets from Core-Shell Sub-Wavelength Spheres. *Opt. Lett.* **2014**, *39*, 3989–3992. [CrossRef]
5. Lee, S.; Li, L.; Wang, Z. Optical Resonances in Microsphere Photonic Nanojets. *J. Opt.* **2014**, *16*, 1–6. [CrossRef]
6. Geints, Y.E.; Panina, E.K.; Zemlyanov, A.A. Control over Parameters of Photonic Nanojets of Dielectric Microspheres. *Opt. Commun.* **2010**, *283*, 4775–4781. [CrossRef]
7. Guo, H.; Han, Y.; Weng, X.; Zhao, Y.; Sui, G.; Wang, Y.; Zhuang, S. Near-Field Focusing of the Dielectric Microsphere with Wavelength Scale Radius. *Opt. Express* **2013**, *21*, 2434–2443. [CrossRef] [PubMed]
8. Devilez, A.; Stout, B.; Bonod, N.; Popov, E. Spectral Analysis of Three-Dimensional Photonic Jets. *Opt. Express* **2008**, *16*, 14200–14212. [CrossRef]
9. Itagi, A.V.; Challener, W.A. Optics of Photonic Nanojets. *J. Opt. Soc. Am. A* **2005**, *22*, 2847–2858. [CrossRef]
10. Luk'Yanchuk, B.S.; Paniagua-Domínguez, R.; Minin, I.; Minin, O.; Wang, Z. Refractive index less than two: photonic nanojets yesterday, today and tomorrow. *Opt. Mater. Express* **2017**, *7*, 1820–1847. [CrossRef]
11. Minin, I.V.; Minin, O.V.; Geints, Y.E. Localized EM and Photonic Jets from Non-Spherical and Non-Symmetrical Dielectric Mesoscale Objects: Brief Review: Localized EM and Photonic Jets from Non-Spherical and Non-Symmetrical. *Ann. Der Phys.* **2015**, *527*, 491–497. [CrossRef]
12. Arya, A.; Laha, R.; Das, G.M.; Dantham, V.R. Enhancement of Raman Scattering Signal Using Photonic Nanojet of Portable and Reusable Single Microstructures. *J. Raman Spectrosc.* **2018**, *49*, 897–902. [CrossRef]
13. Yi, K.J.; Wang, H.; Lu, Y.F.; Yang, Z.Y. Enhanced Raman Scattering by Self-Assembled Silica Spherical Microparticles. *J. Appl. Phys.* **2007**, *101*, 1–6. [CrossRef]
14. Das, G.M.; Ringne, A.B.; Dantham, V.R.; Easwaran, R.K.; Laha, R. Numerical Investigations on Photonic Nanojet Mediated Surface Enhanced Raman Scattering and Fluorescence Techniques. *Opt. Express* **2017**, *25*, 19822–19831. [CrossRef]
15. Lecler, S.; Haacke, S.; Lecong, N.; Crégut, O.; Rehspringer, J.-L.; Hirlimann, C. Photonic Jet Driven Non-Linear Optics: Example of Two-Photon Fluorescence Enhancement by Dielectric Microspheres. *Opt. Express* **2007**, *15*, 4935–4941. [CrossRef]

16. Aouani, H.; Deiss, F.; Wenger, J.; Ferrand, P.; Sojic, N.; Rigneault, H. Optical-Fiber-Microsphere for Remote Fluorescence Correlation Spectroscopy. *Opt. Express* **2009**, *17*, 19085–19092. [CrossRef]

17. Kong, S.-C.; Sahakian, A.V.; Heifetz, A.; Taflove, A.; Backman, V. Robust Detection of Deeply Subwavelength Pits in Simulated Optical Data-Storage Disks Using Photonic Jets. *Appl. Phys. Lett.* **2008**, *92*, 1–3. [CrossRef]

18. Kong, S.-C.; Sahakian, A.; Taflove, A.; Backman, V. Photonic Nanojet-Enabled Optical Data Storage. *Opt. Express* **2008**, *16*, 13713–13719. [CrossRef] [PubMed]

19. Mcleod, E.; Arnold, C.B. Subwavelength Direct-Write Nanopatterning Using Optically Trapped Microspheres. *Nat. Nanotechnol.* **2008**, *3*, 413–417. [CrossRef]

20. Abdurrochman, A.; Lecler, S.; Mermet, F.; Tumbelaka, B.Y.; Serio, B.; Fontaine, J. Photonic Jet Breakthrough for Direct Laser Microetching Using Nanosecond Near-Infrared Laser. *Appl. Opt.* **2014**, *53*, 7202–7206. [CrossRef] [PubMed]

21. Pierron, R.; Chabrol, G.; Roques, S.; Pfeiffer, P.; Yehouessi, J.-P.; Bouwmans, G.; Lecler, S. Large mode area optical fiber for photonic nanojet generation. *Opt. Lett.* **2019**, *44*, 2474–2477. [CrossRef] [PubMed]

22. Surdo, S.; Diaspro, A. Nanopatterning with Photonic Nanojets: Review and Perspectives in Biomedical Research. *Micromachines* **2021**, *12*, 25. [CrossRef]

23. Xudong C.; Erni, D.; Hafner C. Optical Forces on Metallic Nanoparticles Induced by a Photonic Nanojet. *Opt. Express* **2008**, *16*, 13560.

24. Darafsheh, A. Photonic Nanojets and Their Applications. *J. Phys. Photonics* **2021**, *3*, 022001. [CrossRef]

25. Ounnas, B.; Sauviac, B.; Takakura, Y.; Lecler, S.; Bayard, B.; Robert, S. Single and Dual Photonic Jets and Corresponding Backscattering Enhancement With Tipped Waveguides: Direct Observation at Microwave Frequencies. *IEEE Trans. Antennas Propag.* **2015**, *63*, 5612–5618. [CrossRef]

26. Ghenuche, P.; Rigneault, H.; Wenger, J. Hollow-Core Photonic Crystal Fiber Probe for Remote Fluorescence Sensing with Single Molecule Sensitivity. *Opt. Express* **2012**, *20*, 28379–28387. [CrossRef]

27. Allen, K.W.; Kosolapov, A.F.; Kolyadin, A.N.; Pryamikov, A.D.; Mojaverian, N.; Limberopoulos, N.I.; Astratov, V.N. Photonic Jets Produced by Microspheres Integrated with Hollow-Core Fibers for Ultraprecise Laser Surgery. In Proceedings of the 15th International Conference on Transparent Optical Networks (ICTON), Cartagena, Spain, 23–27 June 2013; pp. 1–4.

28. Chabrol, G. Core preservation in single mode optical fibre tip shaping by chemical etching for photonic nanojet material laser processing. *Proc. SPIE* **2018**, *4*, 10681.

29. Minin, I.V.; Minin, O.V.; Liu, Y.-Y.; Tuchin, V.V.; Liu, C.-Y. Concept of Photonic Hook Scalpel Generated by Shaped Fiber Tip with Asymmetric Radiation. *J. Biophotonics* **2020**, *14*, e202000342. [CrossRef] [PubMed]

30. Han, L.; Han, Y.; Wang, J.; Gouesbet, G.; Gréhan, G. Controllable and Enhanced Photonic Jet Generated by Fiber Combined with Spheroid. *Opt. Lett.* **2014**, *39*, 1585–1588. [CrossRef]

31. Zelgowski, J.; Abdurrochman, A.; Mermet, F.; Pfeiffer, P.; Fontaine, J.; Lecler, S. Photonic Jet Subwavelength Etching Using a Shaped Optical Fiber Tip. *Opt. Lett.* **2016**, *41*, 2073–2076. [CrossRef]

32. Cohen, L.G.; Schneider, M.V. Microlenses for Coupling Junction Lasers to Optical Fibers. *Appl. Opt.* **1974**, *13*, 89–94. [CrossRef]

33. Edwards, C.A.; Presby, H.M. Coupling-Sensitivity Comparison of Hemispheric and Hyperbolic Microlenses. *Appl. Opt.* **1993**, *32*, 1573–1577. [CrossRef]

34. Kato, D. Light Coupling from a Stripe-geometry GaAs Diode Laser into an Optical Fiber with Spherical End. *J. Appl. Phys.* **1973**, *44*, 2756–2758. [CrossRef]

35. Lee, K.S.; Barnes, F.S. Microlenses on the End of Single-Mode Optical Fibers for Laser Applications. *Appl. Opt.* **1985**, *24*, 3134–3139. [CrossRef]

36. Eisenstein, G.; Vitello, D. Chemically Etched Conical Microlenses for Coupling Single-Mode Lasers into Single-Mode Fibers. *Appl. Opt.* **1982**, *21*, 3470–3474. [CrossRef] [PubMed]

37. Ryu, S.Y.; Choi, H.Y.; Na, J.; Choi, W.J.; Lee, B.H. Lensed Fiber Probes Designed as an Alternative to Bulk Probes in Optical Coherence Tomography. *Appl. Opt.* **2008**, *47*, 1510–1516. [CrossRef]

38. Zeng, X.H.; Plain, J.; Jradi, S.; Goud, P.R.; Deturche, R.; Royer, P.; Bachelot, R. High Speed Sub-Micrometric Microscopy Using Optical Polymer Microlens. *Chin. Opt. Lett.* **2009**, *7*, 901–903.

39. Zaboub, M.; Guessoum, A.; Demagh, N.-E.; Guermat, A. Fabrication of Polymer Microlenses on Single Mode Optical Fibers for Light Coupling. *Opt. Commun.* **2016**, *366*, 122–126. [CrossRef]

40. Malki, A.; Bachelot, R.; Lauwe, F.V. Two-Step Process for Micro-Lens-Fibre Fabrication Using a Continuous CO_2 Laser Source. *J. Opt. A Pure Appl. Opt.* **2001**, *3*, 291–295. [CrossRef]

41. Choi, H.Y.; Ryu, S.Y.; Na, J.; Lee, B.H.; Sohn, I.-B.; Noh, Y.-C.; Lee, J. Single-Body Lensed Photonic Crystal Fibers as Side-Viewing Probes for Optical Imaging Systems. *Opt. Lett.* **2008**, *33*, 34–36. [CrossRef]

Communication

Direct Writing of Silicon Oxide Nanopatterns Using Photonic Nanojets

Hao Luo [1,2,3], Haibo Yu [1,2,*], Yangdong Wen [1,2], Jianchen Zheng [1,2,3], Xiaoduo Wang [1,2] and Lianqing Liu [1,2,*]

[1] State Key Laboratory of Robotics, Shenyang Institute of Automation, Chinese Academy of Sciences, Shenyang 110016, China; luohao@sia.cn (H.L.); wenyangdong@sia.cn (Y.W.); zhengjianchen@sia.cn (J.Z.); wangxiaoduo@sia.cn (X.W.)

[2] Institutes for Robotics and Intelligent Manufacturing, Chinese Academy of Sciences, Shenyang 110016, China

[3] University of Chinese Academy of Sciences, Beijing 100049, China

* Correspondence: yuhaibo@sia.cn (H.Y.); lqliu@sia.cn (L.L.)

Abstract: The ability to create controllable patterns of micro- and nanostructures on the surface of bulk silicon has widespread application potential. In particular, the direct writing of silicon oxide patterns on silicon via femtosecond laser-induced silicon amorphization has attracted considerable attention owing to its simplicity and high efficiency. However, the direct writing of nanoscale resolution is challenging due to the optical diffraction effect. In this study, we propose a highly efficient, one-step method for preparing silicon oxide nanopatterns on silicon. The proposed method combines femtosecond laser-induced silicon amorphization with a subwavelength-scale beam waist of photonic nanojets. We demonstrate the direct writing of arbitrary nanopatterns via contactless scanning, achieving patterns with a minimum feature size of 310 nm and a height of 120 nm. The proposed method shows potential for the fabrication of multifunctional surfaces, silicon-based chips, and silicon photonics.

Keywords: femtosecond laser processing; photonics nanojet; nanometer-scale pattern; semiconductor materials

Citation: Luo, H.; Yu, H.; Wen, Y.; Zheng, J.; Wang, X.; Liu, L. Direct Writing of Silicon Oxide Nanopatterns Using Photonic Nanojets. *Photonics* **2021**, *8*, 152. https://doi.org/10.3390/photonics8050152

Received: 29 March 2021
Accepted: 29 April 2021
Published: 3 May 2021

Publisher's Note: MDPI stays neutral with regard to jurisdictional claims in published maps and institutional affiliations.

1. Introduction

Silicon is a popular material for multiple applications involving photonics [1,2], integrated circuits [3,4], and silicon-based chips [5]. In recent decades, femtosecond lasers have proven to be a promising tool for processing silicon [6–10], reducing manufacturing costs, increasing efficiency, and enabling large-area processing. The contemporary femtosecond laser treatment of silicon surfaces is focused primarily on removing material using an energy higher than that required for silicon ablation threshold to ablate the material in the processing area. Ablation technology can be used to alter surface reflectivity [11,12] and wettability [13,14] as well as manufacture micro-lens arrays [15,16]. Moreover, owing to the heat accumulation effect, femtosecond lasers with high repetition rates can directly pattern silicon oxide patterns on a silicon substrate, which has potential applications in maskless lithography [17] and microfluidics [18].

Methods of breaking the optical diffraction limit should be developed to increase the resolution of femtosecond laser processing. It has been reported that irradiating a microsphere with parallel-polarized light generates a photonic nanojet (PNJ) in the backlight of the microsphere, which has a full width at half-maximum (FWHM) beyond the optical diffraction limit [19–22], thereby providing a new strategy for femtosecond laser super-resolution processing. A recent study showed that PNJs and femtosecond laser ablation can be combined to improve processing resolution [23–25]. Most studies have focused on ablation using PNJs [26]. However, no studies have reported the use of a combination of PNJs and thermal oxidation, which can be applied to directly manufacture silicon oxide nanopatterns on silicon substrates.

In this study, we combined PNJs and silicon amorphization induced by the heat accumulation in a high-repetition-rate femtosecond laser to enhance processing resolution. We directly manufactured silicon oxide nanopatterns (nanodots and nanopatterns with an aspect ratio of 1:2.5 and a characteristic size of 310 nm) on a silicon substrate and analyzed the oxidation of the silicon; the amorphization of silicon was observed after ablation. Finite-difference time-domain (FDTD) software was used to compute the light field distribution as the plane wave passed through the microsphere; the results were consistent with experimental measurements. Thus, this study demonstrated significant improvements in the machining applicability of femtosecond lasers.

2. Materials and Methods

As shown in Figure 1a, the laser incident on the microsphere formed a PNJ upon emerging at the bottom of the microsphere. To avoid microsphere contamination by debris produced during oxidation, we used low-refractive-index ($n = 1.47$) silica microspheres (diameter = 22 μm, BaseLine, Beijing, China), which ensured that the generated PNJ was focused sufficiently far away from the microspheres. The microspheres were attached to the cantilever of an atomic force microscope (AFM; MLCT, Bruker, Germany) using UV-curable adhesive (NOA63, Edmund Optics) and irradiated for 30 min to fix the attachment (Figure 1b,c) [27]. The low Young's modulus of the MLCT probe ensured that the microspheres and the cantilever were not damaged when the probe touched the surface of the substrate during processing. The AFM probe and the sample were placed on piezo-electric ceramic transducers to accurately control the distance between the sample and the microsphere to an accuracy of 0.4 nm.

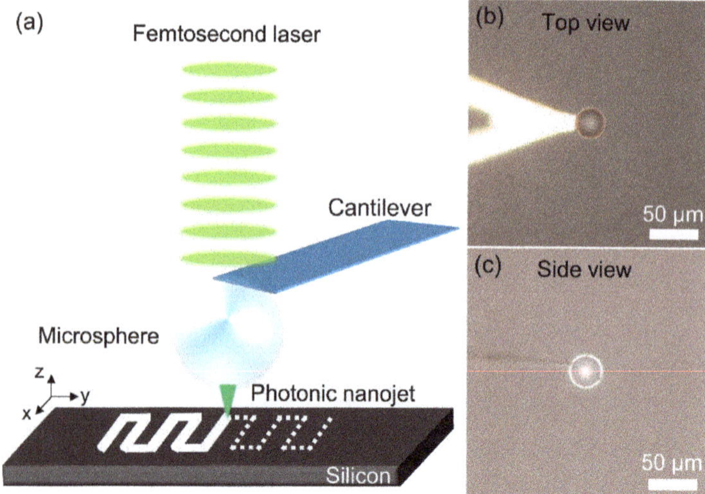

Figure 1. (a) Schematic showing the direct writing of silicon oxide patterns on a silicon substrate via a PNJ. (b) Top and (c) side views of a microsphere attached to the cantilever of an AFM. The diameter of the fused silica microspheres ($n = 1.47$) is approximately 22 μm.

The laser beam was focused through an objective lens (NA = 0.3, MPlanFLN, Olympus) and irradiated onto the microsphere to maximize the laser energy. We used a femtosecond laser (PHAROS-6W, Light Conversion) with a wavelength of 515 nm, pulse duration of 290 fs, and a maximum repetition rate of 200 kHz. The high repetition rate facilitated heat accumulation on the silicon surface, thus accelerating the oxidation. In this experiment, the repetition frequency was fixed at 200 kHz, and the scanning speed was 10 μm/s.

We used Si<100> substrates, which were cleaned with a solution of ammonia, hydrogen peroxide, and ultrapure water at a ratio of 1:1:5. In addition, ultrasonic treatment was performed for 5 min at 50 °C to remove impurities generated during processing. All experimental results were obtained via AFM (Dimension Icon, Bruker) or scanning electron microscopy (SEM; Zeiss EVO MA10). Prior to SEM characterization, the samples were coated with a gold layer (approximate thickness of 30 nm) using a magnetron sputtering instrument.

In addition to the experimental characterization, optical field distribution was simulated using the FDTD computational technique. The simulations were performed for an area of 60 μm × 30 μm, mesh size of $\lambda/100$, and the boundary matching layers were set as perfect absorption layers to ensure accuracy. The incident light was a plane wave with a wavelength of 515 nm. The refractive indices of the microsphere (diameter = 22 μm) and the background medium were set as 1.47 and 1, respectively. The simulation conditions were consistent with those of the experiment.

3. Results and Discussion

3.1. Process of Silicon Oxide Microstructure Induced by PNJ

A convex structure, diameter of approximately 2.9 μm was achieved by directly focusing a 200 kHz femtosecond laser (Figure 2a). Energy-dispersive spectroscopy (EDS) was used to analyze the constitution of the convex structure, thus showing that a femtosecond laser with a high repetition rate can oxidize silicon to silicon oxide (Figure 2b) [17,28,29]. In order to improve the manufacturing resolution, microspheres were used to generate PNJ.

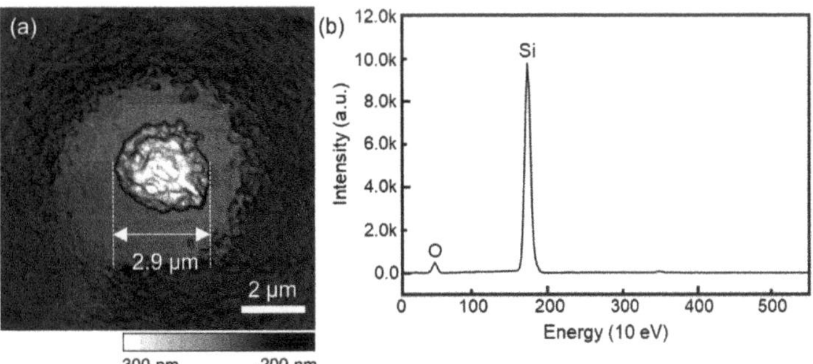

Figure 2. (**a**) High-repetition-rate femtosecond laser-induced convex structures, directly focused by an object lens (NA = 0.3, MPlanFLN, Olympus). (**b**) EDS of convex structures showing silicon was oxidized by a femtosecond laser.

Figure 3a shows a schematic of the oxidation of the silicon substrate by the PNJ. Owing to the extremely high energy density of the PNJ, the silicon substrate was ablated, forming micro/nanocavities [24]. After the formation of the micro-/nanocavities, there was not sufficient space for the heat accumulated to dissipate because of the high-repetition laser pulses, causing the temperature in the pores to be higher than the temperature oxidating the silicon substrate. The resulting oxidation converted the silicon to silica. In addition, because the ablation threshold of silicon is much lower than that of silicon dioxide [30,31], convex silicon oxide formed on the silicon substrate (Figure 3b).

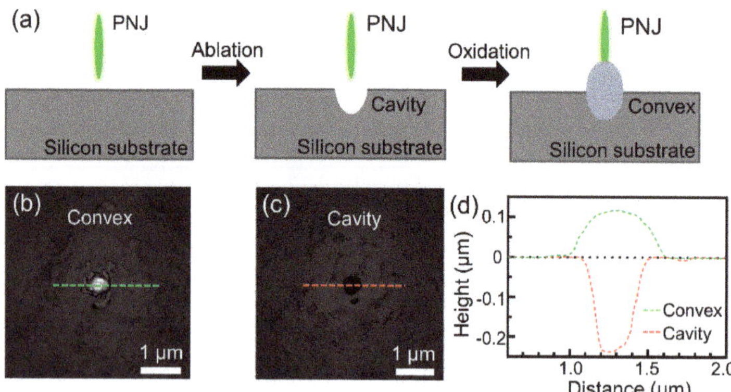

Figure 3. (**a**) Oxidation of the silicon substrate by the PNJ. (**b**) AFM image of convex silicon oxide induced by the PNJ. (**c**) AFM image showing the cavity etched from silicon oxide on silicon with 10% HF. (**d**) Cross-sectional views of the dotted lines in (**b**,**c**).

Hydrofluoric acid (HF) is strongly corrosive to silicon oxides; therefore, we used 10% HF to remove the convex oxide. This process created a cavity with an FWHM of 265 nm on the silicon substrate (Figure 3c), confirming that the main component of the convex structure induced by the high-repetition-rate femtosecond laser is silicon oxide.

The morphologies of the convex structure and the cavity were analyzed using AFM, as shown in Figure 3d. The convex structure has an FWHM of 310 nm, and the ratio of height to width is approximately 1:2.5. Note that the convex structure was formed using direct laser irradiation without introducing an etching reagent. Compared without microspheres (Figure 2a), the character size of the manufactured features is greater by an order of magnitude.

3.2. Simulation Analysis of PNJ

To analyze the PNJ processing mechanism in greater detail, we conducted a simulation using the experimental conditions to calculate the light field distribution generated by the femtosecond laser as it passes through the microsphere (Figure 4). When the PNJ emerges from the microsphere, the local light intensity is significantly enhanced (Figure 4a). The energy distribution of the light field along the z-axis is shown in Figure 4b. Starting from the direction of incidence (indicated by the green arrows in Figure 4a), the energy of the light field oscillates violently because of the interference between the reflected and incident light [32]. Outside the microsphere, the energy of the light field increases rapidly and is the highest at a distance of 4.5 µm from the microsphere, at which point the energy intensity is approximately 30 times greater than that of the incident light energy. However, the energy decreases rapidly as the distance continues to increase. The length of the PNJ is approximately 5 µm, exceeding the focal spot length of the high-NA objective lens, which has a longer working distance. In addition, we analyzed the energy profile in the xy-plane at the maximum light intensity, as shown in Figure 4c,d. A side lobe can be observed near the PNJ, which has an intensity equal to approximately one-third that of the PNJ. The FWHM of the PNJ is approximately 400 nm (Figure 4d), which corresponds to the size of the machined protrusion.

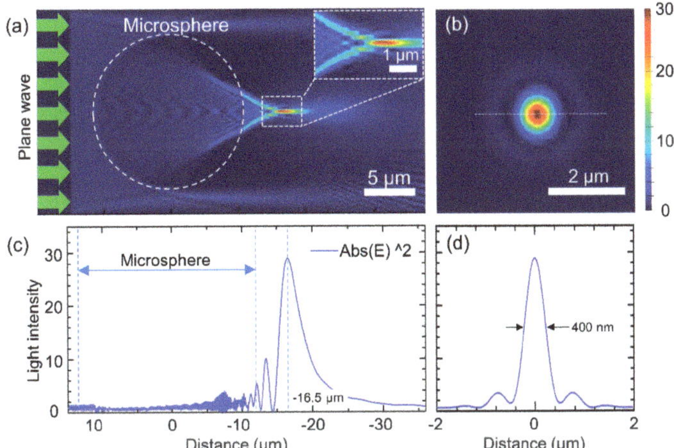

Figure 4. FDTD simulation of the light field distribution of the PNJ. (**a**) Setup of the PNJ simulation. The incident light intensity was normalized, with the color bar representing the scaled light intensity. The inset shows the magnified PNJ. (**b**) The light field distribution in the xy-plane of the PNJ. (**c**) The light field distribution along the z-axis of the PNJ. The horizontal axis represents distance, and the vertical axis represents the energy intensity of the light field. The center of the microsphere coincides with the zero point of the horizontal axis. (**d**) Prolife centered about the dotted line in (**b**).

3.3. Effect of Laser Power on the Size of the Oxidized Region

Energy, as an important parameter in laser processing, determines the size of the machined structures. The relationship between the energy density of the PNJ and the size of the oxidation region for femtosecond lasers of different powers is shown in Figure 5. Specifically, Figure 5a–d show SEM images corresponding to laser powers of 158, 211, 277, and 331 µW, respectively. The protruding surface was relatively rough, owing to uneven oxidation. In part, this is due to the accumulation of oxidized debris generated during the ablation. As in ordinary laser processing, a higher energy density results in larger feature sizes (Figure 5e). The minimum feature size obtained during the experiment was 290 nm. Owing to the nonlinear absorption of silicon and ablation fluence threshold [33], the FWHM of the convex structure was smaller than the simulated value, which is consistent with the simulation results. In addition, Figure 4b shows that the energy distribution at the peak PNJ intensity is approximately Gaussian. Therefore, by combining accurate laser energy control with the nonlinear absorption of silicon, high-resolution silica protrusions can be achieved and used to create patterns.

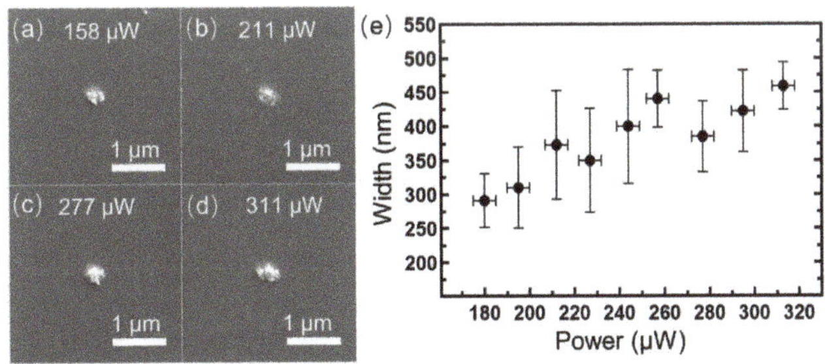

Figure 5. Influence of energy density on the machined feature size with microspheres. (**a–d**) SEM images of silicon substrates irradiated with lasers with powers of 158, 211, 277, and 331 µW, respectively. (**e**) Width of the oxidation-induced convex silicon oxide structures as a function of laser power.

3.4. Directly Writing Arbitrary Silicon Dioxide Pattern

Finally, we applied the one-step process to realize direct silica patterning by scanning. The position of the microsphere was maintained, while the piezoelectric ceramic transducer was programmed to trace an expected path. Note that the substrate was separated from the microspheres by an approximate distance of 4.5 μm, with this noncontact processing method protecting the microspheres from surface splashes. The pattern shown in the SEM image in Figure 6 was produced using a laser power of 158 μW, a scanning speed of 10 μm/s, and a laser repetition frequency of 200 kHz. We achieved a silicon oxide pattern preparation of 100 μm × 100 μm (Figure 6a), and the width of the prepared features was approximately 350 nm (Figure 6c). Owing to the nonuniformity of oxidation, achieving highly uniform patterns via this approach is difficult. Nevertheless, it may be possible to reduce the roughness of silicon dioxide nanowires by using a femtosecond laser with a higher repetition rate.

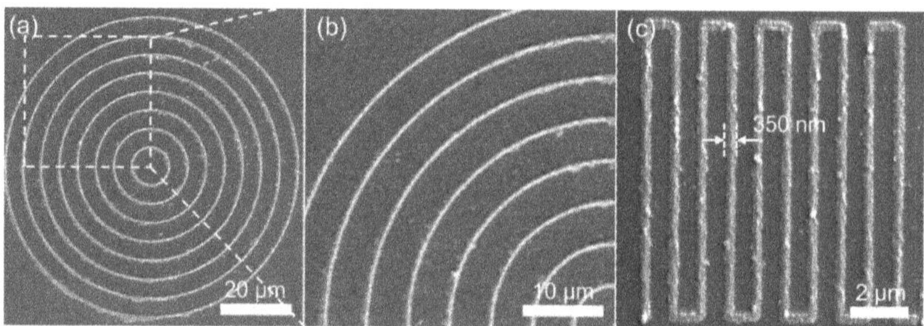

Figure 6. SEM image of silicon oxide nanopatterns, (**a**,**b**) concentric circles and (**c**) grid.

4. Conclusions

This study demonstrates the direct writing of silicon oxide nanopatterns on a silicon substrate with an FWHM of approximately 350 μm and an aspect ratio of 1:2.5. By using the FWHM of the PNJ to overcome the optical diffraction limit, along with using a high-repetition-rate femtosecond laser, the patterning of silica nanostructures on a silicon substrate was achieved via a one-step process. The minimum size of the prepared nanopattern is approximately 310 nm. With further refinement, we believe that the combination of nonlinear material absorption and the precise laser energy control can be used to fabricate silica patterns with a resolution <300 nm or better. As such, the approach described herein has the potential to be utilized for applications requiring silicon-based chips and optical waveguides.

Author Contributions: L.L., H.L. and H.Y. conceived the basic idea of this study and designed its experiments; H.L. and X.W. designed the optical system and performed the experiments; H.L., H.Y. and L.L. analyzed the data and co-wrote this paper; L.L. and H.Y. supervised the project; J.Z. and Y.W. assisted in the experimental processes. All authors contributed to the general discussion and revision of the manuscript. All authors have read and agreed to the published version of the manuscript.

Funding: This research was funded by National Natural Science Foundation of China (grant numbers 61925307, 61727811, 61973298, 61803366 and 61821005); External Cooperation Program of the Chinese Academy of Sciences (grant number 173321KYSB20170015); CAS Interdisciplinary Innovation Team (JCTD-2019-09); Liaoning Revitalization Talents Program (grant number XLYC1807006); Youth Innovation Promotion Association of the Chinese Academy of Sciences (grant number Y201943).

Data Availability Statement: The data are available from the corresponding author upon reasonable request.

Conflicts of Interest: The authors declare no conflict of interest.

References

1. Priolo, F.; Gregorkiewicz, T.; Galli, M.; Krauss, T.F. Silicon nanostructures for photonics and photovoltaics. *Nat. Nanotechnol.* **2014**, *9*, 19–32. [CrossRef] [PubMed]
2. Wu, X.; Pan, X.; Quan, B.; Wang, L. Optical modulation of terahertz behavior in silicon with structured surfaces. *Appl. Phys. Lett.* **2013**, *103*, 121112. [CrossRef]
3. Wang, C.G.; Wu, X.Z.; Di, D.; Dong, P.T.; Xiao, R.; Wang, S.Q. Orientation-dependent nanostructure arrays based on anisotropic silicon wet-etching for repeatable surface-enhanced Raman scattering. *Nanoscale* **2016**, *8*, 4672–4680. [CrossRef]
4. Lin, C.; Huang, N.; Povinelli, M.L. Effect of aperiodicity on the broadband reflection of silicon nanorod structures for photovoltaics. *Opt. Express* **2012**, *20*, A125–A132. [CrossRef] [PubMed]
5. Andersson, H.; Berg, A. Microfabrication and microfluidics for tissue engineering: State of the art and future opportunities. *Lab Chip* **2004**, *4*, 98–103. [CrossRef]
6. Malinauskas, M.; Žukauskas, A.; Hasegawa, S.; Hayasaki, Y.; Mizeikis, V.; Buividas, R.; Juodkazis, S. Ultrafast laser processing of materials: From science to industry. *Light Sci. Appl.* **2016**, *5*, e16133. [CrossRef]
7. Sugioka, K.; Cheng, Y. Ultrafast lasers—Reliable tools for advanced materials processing. *Light Sci. Appl.* **2014**, *3*, e149. [CrossRef]
8. Chen, T.; Wang, W.; Tao, T.; Pan, A.; Mei, X. Multi-scale micro-nano structures prepared by laser cleaning assisted laser ablation for broadband ultralow reflectivity silicon surfaces in ambient air. *Appl. Surf. Sci.* **2020**, *509*, 145182. [CrossRef]
9. Vorobyev, A.Y.; Guo, C. Multifunctional surfaces produced by femtosecond laser pulses. *J. Appl. Phys.* **2015**, *117*, 033103. [CrossRef]
10. Chicbkov, B.N.; Momma, C.; Nolte, S. Femtosecond, picosecond and nanosecond laser ablation of solids. *Appl. Phys. A* **1996**, *63*, 109–115. [CrossRef]
11. Zou, T.; Zhao, B.; Xin, W.; Wang, Y.; Wang, B.; Zheng, X.; Xie, H.; Zhang, Z.; Yang, J.; Guo, C. High-speed femtosecond laser plasmonic lithography and reduction of graphene oxide for anisotropic photoresponse. *Light Sci. Appl.* **2020**, *9*, 69. [CrossRef]
12. Chen, T.; Wang, W.; Tao, T.; Pan, A.; Mei, X. Broad-Band Ultra-Low-Reflectivity Multiscale Micro–Nano Structures by the Combination of Femtosecond Laser Ablation and In Situ Deposition. *ACS Appl. Mater. Interfaces* **2020**, *12*, 49265–49274. [CrossRef]
13. Huo, J.; Yang, Q.; Yong, J.; Fan, P.; Lu, Y.; Hou, X.; Chen, F. Bubble Passage: Underwater Superaerophobicity/Superaerophilicity and Unidirectional Bubble Passage Based on the Femtosecond Laser-Structured Stainless Steel Mesh. *Adv. Mater. Interfaces* **2020**, *7*, 2070077. [CrossRef]
14. Zhan, Z.; ElKabbash, M.; Cheng, J.; Zhang, J.; Singh, S.; Guo, C. Highly Floatable Superhydrophobic Metallic Assembly for Aquatic Applications. *ACS Appl. Mater. Interfaces* **2019**, *11*, 48512–48517. [CrossRef] [PubMed]
15. Bian, H.; Wei, Y.; Yang, Q.; Chen, F.; Zhang, F.; Du, G.; Yong, J.; Hou, X. Direct fabrication of compound-eye microlens array on curved surfaces by a facile femtosecond laser enhanced wet etching process. *Appl. Phys. Lett.* **2016**, *109*, 221109. [CrossRef]
16. Cao, J.J.; Hou, Z.S.; Tian, Z.N.; Hua, J.G.; Zhang, Y.L.; Chen, Q.D. Bioinspired Zoom Compound Eyes Enable Variable-Focus Imaging. *ACS Appl. Mater. Interfaces* **2020**, *12*, 10107–10117. [CrossRef] [PubMed]
17. Liu, X.Q.; Yu, L.; Ma, Z.C.; Chen, Q.D. Silicon three-dimensional structures fabricated by femtosecond laser modification with dry etching. *Appl. Opt.* **2017**, *56*, 2157–2161. [CrossRef]
18. Kumar, K.; Lee, K.K.C.; Herman, P.R.; Nogami, J.; Kherani, N.P. Femtosecond laser direct hard mask writing for selective facile micron-scale inverted-pyramid patterning of silicon. *Appl. Phys. Lett.* **2012**, *101*, 222106. [CrossRef]
19. Ferrand, P.; Wenger, J.; Devilez, A.; Pianta, M.; Stout, B.; Bonod, N.; Popov, E.; Rigneault, H. Direct imaging of photonic nanojets. *Opt. Express* **2008**, *16*, 6930–6940. [CrossRef] [PubMed]
20. Zhu, J.; Goddard, L.L. Spatial control of photonic nanojets. *Opt. Express* **2016**, *24*, 30444–30464. [CrossRef]
21. Kim, M.S.; Lahijani, B.V.; Descharmes, N.; Straubel, J.; Negredo, F.; Rockstuhl, C.; Hayrinen, M.; Kuittinen, M.; Roussey, M.; Herzig, H.P. Subwavelength Focusing of Bloch Surface Waves. *ACS Photonics* **2017**, *4*, 1477–1483. [CrossRef]
22. Wang, Z.; Guo, W.; Li, L.; Luk, B.; Khan, A.; Liu, Z.; Chen, Z.; Hong, M. Optical virtual imaging at 50 nm lateral resolution with a white-light nanoscope. *Nat. Commun.* **2011**, *2*, 218. [CrossRef] [PubMed]
23. Wen, Y.; Wang, F.; Yu, H.; Li, P.; Liu, L.; Li, W.J. Laser-nanomachining by microsphere induced photonic nanojet. *Sens. Actuators A Phys.* **2017**, *258*, 115–122. [CrossRef]
24. Yan, B.; Yue, L.; Monks, J.N.; Yang, X.; Xiong, D.; Jiang, C.; Wang, Z. Superlensing plano-convex-microsphere (PCM) lens for direct laser nano-marking and beyond. *Opt. Lett.* **2020**, *45*, 1168–1171. [CrossRef] [PubMed]
25. Jacassi, A.; Tantussi, F.; Dipalo, M.; Biagini, C.; Maccaferri, N.; Bozzola, A.; Angelis, F.D. Scanning Probe Photonic Nanojet Lithography. *ACS Appl. Mater. Interfaces* **2017**, *9*, 32386–32393. [CrossRef] [PubMed]
26. Darafsheh, A. Photonic nanojets and their applications. *J. Phys. Photonics* **2021**, *3*, 022001. [CrossRef]
27. Wang, F.; Liu, L.; Yu, H.; Wen, Y.; Yu, P.; Liu, Z.; Wang, Y.; Li, W.J. Scanning superlens microscopy for non-invasive large field-of-view visible light nanoscale imaging. *Nat. Commun.* **2016**, *7*, 13748. [CrossRef]
28. Kiani, A.; Venkatakrishnan, K.; Tan, B. Direct laser writing of amorphous silicon on Si-substrate induced by high repetition femtosecond pulses. *J. Appl. Phys.* **2010**, *108*, 074907. [CrossRef]
29. Kiani, A.; Venkatakrishnan, K.; Tan, B. Direct patterning of silicon oxide on Si-substrate induced by femtosecond laser. *Opt. Express* **2010**, *18*, 1872–1878. [CrossRef]

30. Bloembergen, N. Laser-induced electric breakdown in solids. *IEEE J. Quantum Electron.* **1974**, *10*, 375. [CrossRef]
31. Bonse, J.; Baudach, S.; Krüger, J.; Kautek, W.; Lenzner, M. Femtosecond laser ablation of silicon–modification thresholds and morphology. *Appl. Phys. A* **2002**, *74*, 19–25. [CrossRef]
32. Wen, Y.; Yu, H.; Zhao, W.; Wang, F.; Wang, X.; Liu, L.; Li, W.J. Photonic Nanojet Sub-Diffraction Nano-Fabrication With in situ Super-Resolution Imaging. *IEEE Trans. Nanotechnol.* **2019**, *18*, 226–233. [CrossRef]
33. Garcia-Lechuga, M.; Utéza, O.; Sanner, N.; Grojo, D. Evidencing the nonlinearity independence of resolution in femtosecond laser ablation. *Opt. Lett.* **2020**, *45*, 952–955. [CrossRef] [PubMed]

MDPI
St. Alban-Anlage 66
4052 Basel
Switzerland
Tel. +41 61 683 77 34
Fax +41 61 302 89 18
www.mdpi.com

Photonics Editorial Office
E-mail: photonics@mdpi.com
www.mdpi.com/journal/photonics